Exploring the Brazos River

*For Peggy Jackson —
Keep loving the Brazos!*

[signature]

Jerry T. [signature]

RIVER BOOKS

Sponsored by the

 River Systems Institute
at Texas State University

Andrew Sansom, General Editor

*A list of titles in this series is available
at the back of the book.*

A☆M nature guides

TEXAS A&M UNIVERSITY PRESS   COLLEGE STATION

Exploring the Brazos River

FROM BEGINNING TO END

By Jim Kimmel

Photographs by Jerry Touchstone Kimmel

Foreword by Andrew Sansom

The paper used in this book
meets the minimum requirements
of the American National Standard
for Permanence of Paper
for Printed Library Materials,
z39.48–1984.
Binding materials have been
chosen for durability.

LIBRARY OF CONGRESS
CATALOGING-IN-PUBLICATION DATA

Kimmel, Jim, 1943–
    Exploring the Brazos River : from beginning to end / by Jim
Kimmel ; photographs by Jerry Touchstone Kimmel ; foreword
by Andrew Sansom. — 1st ed.
        p. cm. — (River books)
    Includes bibliographical references and index.
    ISBN-13: 978-1-60344-432-3 (flexbound : alk. paper)
    ISBN-10: 1-60344-432-7 (flexbound : alk. paper)
    ISBN-13: 978-1-60344-480-4 (e-book)
    ISBN-10: 1-60344-480-7 (e-book)
    1. Brazos River Valley (Tex.)—Description and travel.
2. Brazos River Valley (Tex.)—History.   3. Stream ecology—
Texas—Brazos River Valley.   4. Brazos River Valley (Tex.)—
Guidebooks.   I. Title.   II. Series: River books (Series).
    F392.B842K56   2011
    976.4—dc22
    2011007987

To Madelyn Kimmel,
Kristi Plymale,
Jerry Touchstone Kimmel,
Jamie Shelton,
& Julie Grabbs

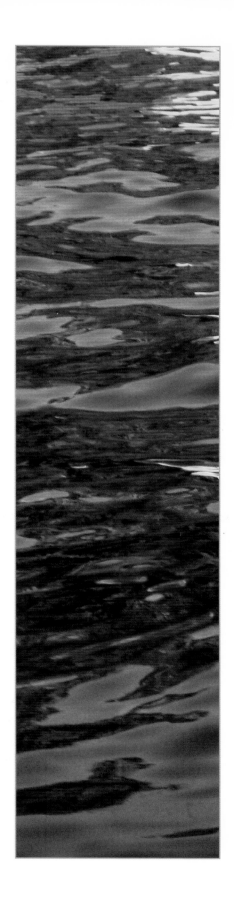

CONTENTS

Foreword, by Andrew Sansom   IX

Acknowledgments   XI

Introduction   1

Chapter 1   *Water Runs Downhill*   7

Chapter 2   *The Brazos as an Ecological System*   27

Chapter 3   *The Lost River*   45

Chapter 4   *Many Arms of God*   58

Chapter 5   *John Graves's Dammed River*   76

Chapter 6   *The (Almost) Free Brazos*   115

Chapter 7   *The Evolving Brazos*   150

Appendix   *Plant and Animal Species of the Brazos River* 155

References   161

Index   173

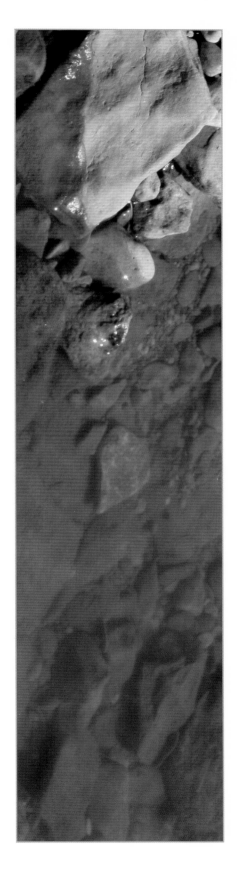

FOREWORD

My earliest memories are of the Brazos. Where I lived, near its mouth on the Texas coast, the river is big and muddy. It seemed huge in my childhood, and we generally stayed away from it because it was intimidating. The river flowed behind the football stadium at Brazosport High School where I was a student, and I will never forget the day I finally mustered up enough courage to sneak away from football practice and swim across it. At that point, the river is several hundred yards wide and by the time I got to the other side, I was half a mile downstream due to the swiftly moving current. That foolish adventure taught me great respect for the Brazos, a respect that has stayed with me all my life, making my personal connection with the river all the more indelible.

Canoeing the river as an adult, I was struck with the fact that people living along it viewed it mostly as a waste disposal site, dumping all manner of trash down its steep banks for the floods to carry away. Yet the Brazos has historically teemed with life, as when the big tarpon rolled in its lower reaches and drew anglers from all over the country.

Few of our Texas rivers have as enduring a connection to the life and times of our state as the Brazos. Though discovered and named by Spanish explorers, the river lies at the very heart of Anglo Texas and is central to many of the most significant places and events of the Texas Revolution and the years that followed. Stephen F. Austin's first boatload of colonists washed ashore after a shipwreck at the river's mouth. Austin's Colony was located along the river and the Texas Declaration of Independence, which was drafted by many of the people he brought here, was signed on its banks. The first capital of the Republic of Texas was at West Columbia in the Brazos bottoms of present-day Brazoria County.

The physical history of the Brazos River is equally compelling. It is the only one of our Texas rivers that flows directly into the Gulf of Mexico, probably because incredible amounts of silt, carried downstream for millions of years, have pushed the frontal mainland at its mouth literally out into open water. This heavy siltation required so much expensive dredging to maintain navigation in the Freeport Harbor Channel, where the Brazos met the sea, that in an amazing and audacious early twentieth-century navigation project, the mouth was actually moved up the coast several

miles, freeing the harbor of silt. The original mouth is today known as the "Old River" and is the heart of one of the great petrochemical complexes of the world.

Farther upstream, stretches of the Brazos maintain much of the river's original natural splendor. One of Texas' great paddling opportunities is the run through Dark Valley below the dam at Possum Kingdom, and above the lake, the Salt, Clear, and Double Mountain Forks are among the most beautiful streams in the nation.

The Brazos' unique blend of natural and cultural history is captured in this ninth volume of River Books by the dedicated and talented husband and wife team of Jim and Jerry Kimmel. Jim, a distinguished geographer at Texas State University, also has a past with the Brazos, having spent part of his life in Waco, and his words reflect both his scholarship and his love. Jerry's stunning photos bring the river to life in a way few others have managed to do. Together, the work here is their second collaboration in our series, following the story of the San Marcos, another of Texas' most iconic streams.

This book would not have been possible without the support of another icon of Texas, Houston's Oshman family, whose roots along the Brazos are deeper than mine. In 1914, an orphan from Russia, Jacob Oshman, arrived in Richmond on the lower Brazos to live with his aunt. At the age of nineteen, the young immigrant opened a small department store with a relative in 1919. This store, Oshman's, became the foundation of one of America's great sporting goods empires. In the early years of his growing retail business, after Jacob Oshman opened the first sporting goods store in Houston, his daughters, Judy and Marilyn, vividly remember crossing the great river on the way back and forth to Richmond. As Marilyn recollects:

The landscape was flat and green with a searing blue sky. There were patches of rice fields under water sometimes, and we liked that because the water broke the monotony until we would get to Sugarland and see the Imperial Sugar sign and building. Then, the scenery would immediately return to the flat land, and we would doze off or play games until we could see the old bridge crossing the Brazos. Daddy would start to tell the story of swimming in the river when he was a boy. He would talk about all the creatures that he saw while he was on its shore. He talked about fishing in the river and catching little fish with worms on cane poles. But most of all what I remember is the sight of the river itself. It always had a dark green color, and the water seemed to always be moving fast. There were many trees and overgrown vines and shrubs at its side. I thought trolls or bayou mojo might be hiding there. The ripples were always there, and the little white waves made the river more enticing.

The hot Texas summers seemed to invite you in. Daddy talked about how dangerous the currents were, but its aura is still something I can never forget. Our mother said we absolutely could not go into that water. I never stepped a foot into the Brazos.

For many years I thought the Brazos sort of belonged to Richmond and my father was like a Tarzan hero because he spent time and swam there. Years later, driving along highway 10, we crossed the Brazos many times. What I didn't realize is that a river that can meander back and forth for so long might really have some of the magical powers that attracted me on those early rides to Richmond. Aren't we lucky that it's still running?

Today, Marilyn and Judy Oshman maintain a lovely preserve in the river's bottomland and together direct the foundation established in their father's name.

The Brazos commands our respect in a way no other river manages to do, and it is interwoven with our culture in a way that touches our collective soul.

—Andrew Sansom

ACKNOWLEDGMENTS

I thank Andrew Sansom, executive director of the River Systems Institute at Texas State University–San Marcos, for his continuing encouragement and for his support for the publication of this book. I am especially grateful to Shannon Davies, Louise Lindsey Merrick Editor for the Natural Environment at Texas A&M University Press, for her patience and encouragement. Connie and John Giesenschlag and Penne and Mark Jackson spent many hours helping Jerry Touchstone Kimmel and me understand agriculture on the Brazos. We value their care for the land and appreciate their hospitality. Doyle Mosier, director of the River Studies Program of the Texas Parks and Wildlife Department, provided very valuable review and comments, as did Ed Lowe and several staff members at the Brazos River Authority. Finally, I thank the anonymous reviewers, whose comments greatly strengthened the text, Maureen Bemko for her careful editing, and Kevin Schwartz for his cartographic skills.

Exploring the Brazos River

Introduction

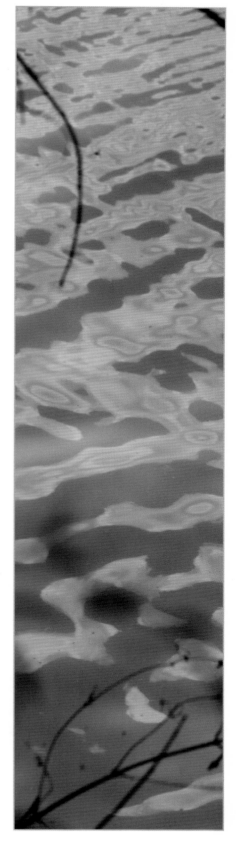

We are flying a small plane west from Abilene to Albuquerque in June. We check flight service weather and learn that a dry line is moving eastward and the forecast is for scattered thunderstorms. We haven't flown here before, but we hope, if there is a storm, it won't amount to much, that we can go around it. Less than an hour after takeoff we see a very dark and wide band of clouds ahead. We watch as they rise quickly, and we realize that the front is too big and moving too fast for us to go around. Soon we hear other pilots on the radio talking about the "Marfa Monster" that is developing, and we ask one what he means. He says that if you don't know what the Marfa Monster is you'd better find a place to land immediately. We land at Lubbock, quickly tie the plane down, and run inside as the storm hits. The wind is 40 mph, driving sheets of rain for forty-five minutes. Another pilot says we were lucky there wasn't the usual hail with the storm, or we would have a seriously damaged airplane.

The storm passes as quickly as it arose, and we head west again, in clear, smooth air. But as we look down it seems that water is everywhere. Even the smallest creeks are running high and red, and we see water over several roads. We have just met the spirit of the Brazos River, a spirit that makes itself known for hundreds of miles through the heart of Texas.

In *Goodbye to a River*, one of the classic river books of North America, John Graves offered two pointed observations about the Brazos River. First, he assumed that only a few people would genuinely care about the river. Second, he said that a person could really know only part of a river, because a river is so many things. This book challenges those observations, not to refute Graves, but out of necessity. Necessity, because fifty years after Graves wrote about his trip down the Brazos it is even more important that we learn to appreciate the Brazos River as a whole and complex entity.

That you can know only part of a river is true in the way John Graves

Torrential rainstorms in the upper Brazos basin, sometimes called Marfa Monsters,
cause rapid runoff resulting in erosion and high flows, thus giving the Brazos its most
notable characteristics—muddiness and frequent high floods.

Introduction

The section of the Brazos that John Graves wrote about in Goodbye to a River *flows through low limestone hills below the dam that forms Possum Kingdom Lake. The dam moderates the flow and removes much of the silt, making this part of the Brazos much clearer than it would be naturally.*

knew his part of the Brazos southwest of Fort Worth—a deep knowing from childhood. My childhood memories are of the Bosque River, an important Brazos tributary. But as a geographer I learned a different way to know a river, a more comprehensive way that recognizes that a river is both product and producer of the land. We can know a river as a whole entity if we look at it that way. We need to understand rivers as the complex systems they are, because our failure to think of rivers holistically has led us to make many mistakes in our relationships with them.

The purpose of this book is to help you understand how the Brazos River "works" and, perhaps even more, to inspire you to *want* to understand how the Brazos and other rivers work—what they do and how and why. I draw on river science and use terms such as cubic feet per second (cfs) to describe flow and total dissolved solids (TDS) and total suspended solids (TSS) to describe water quality. For those of you who steered a course away from science in your education, don't be alarmed. These are just commonsense terms. Think about how solid salt dissolves in water but sand does not.

So come with us to learn about a great Texas river, the longest within the state, and the river with the largest flow in Texas. We will explore the entire river system, including the tributaries, from New Mexico to the Gulf of Mexico. We will cross almost all of its bridges, camp on its banks, and paddle upstream and downstream in our kayaks. We will document the problems and challenges of the Brazos, but we will also look for those places of excitement, beauty, and learning—some of them surprising.

With its topographic drainage beginning on the High Plains in New Mexico and cutting through Texas' geography and history to the Gulf of Mexico, the Brazos River is varied and complex.

CHAPTER 1

Water Runs Downhill

Rivers are one of the most dramatic features of a continent. They are the inevitable result of precipitation falling across the land, coalescing into streams, and uniting into ever larger streams and rivers. Over millions of years, these networks of flowing waters have delivered sediments and nutrients to downstream areas, sometimes eroding valleys and at other times depositing sediments, before eventually reaching the sea or an inland lake.

Arthur C. Benke and Colbert E. Cushing, *Rivers of North America* (2005)

AT ITS MOST BASIC, a river is water running downhill. But a wonderful variety of processes interact in that seemingly simple run to the sea. What are the sources, amounts, and timing of the flow? What are the geologic and ecological characteristics of the land from which the water flows? How have those affected humans, what have been our responses, and how have the river and the land responded to us in this fundamentally serious dance? Most especially, what are the trends and implications for the future?

This chapter and the next examine the context that will enable us to con-sider those questions. Subsequent chapters examine four distinct but related segments of the Brazos River.

Embracing a Big River

In about 1680 a Chinese court painter named Wang Hui painted a mostly imaginary image of the entire four-thousand-mile length of the Yangtze River on a fifty-three-foot-long scroll (Winchester 2004). But today we have something even more amazing. Using the satellite photos available in programs such as Google Earth® we can virtually fly the length of any

The Brazos River Basin and Ecoregions of Texas

Brazos River Basin

① Trans-Pecos	⑥ Cross Timbers and Prairies
② High Plains	⑦ Blackland Prairie
③ Rolling Plains	⑧ Post Oak Savannah
④ Edwards Plateau	⑨ Gulf Prairies and Marshes
⑤ South Texas Plains	⑩ Piney Woods

Data Sources:
The Texas Commission on Environmental Quality,
the Texas Parks & Wildlife Department,
and the United States Geological Survey

Created by: Kevin Schwartz
May, 2009

The four segments of the Brazos, clockwise from upper left: the Lost River that no longer contributes flow; the Many Arms of God, where four tributaries come together to form the Brazos; John Graves's Dammed River, where flow is controlled by various dams; and the (Almost) Free Brazos, from Waco to the Gulf, one of longest stretches of undammed river in the United States.

river. Let's follow the Brazos from where it once flowed sporadically in eastern New Mexico all the way to its mouth at the Gulf of Mexico. You can read the description below, or you can "fly" to specific places on Google Earth® by typing in the name of your destination or by using the longitude and latitude coordinates for specific locations.

Fly to Running Water Draw north of Clovis, New Mexico (34° 35' W, 103° 25' N) to see the uppermost topographic drainage of the Brazos. Head southeast and you won't see anything that looks like a river for almost one hundred miles. The most obvious feature on the landscape is the multitude of circular fields of cotton watered by centerpoint irrigation, drawing from wells drilled into the Ogallala Aquifer. Go to the Muleshoe National Wildlife Refuge and you will see natural and modified playa lakes of the High Plains (33° 58' W, 102° 44' N). The land below is topographically part of the Brazos watershed, but it no longer contributes any water to the river's flow. As you fly over Lubbock, Texas, the land appears to be flat because there is little topographic relief, but it does slope to the southeast. However, a few miles east of Lubbock the landscape becomes much rougher after you cross the Caprock Escarpment. The land surface is three hundred to four hundred feet lower and cut by erosion. Zoom in at 33° 36' N, 101° 42' W and look at Buffalo Springs Lake and Ransom Can-

The town of Mineral Wells was not actually built beside the Brazos, but its access to the south was cut off by the river. This high bridge over the Brazos solved that problem and became a notable place for taking family photographs.

yon Lake. Follow the rugged canyon southeast to 33° 03' N, 101° 04' W to see the new Lake Alan Henry on the Double Mountain Fork of the Brazos. This is the first major reservoir in the Brazos system. Move up to 33° 21' 24" N and 100° 30' 02" W and look at the salt flats at the head of the Salt Fork of the Brazos.

Fly to Possum Kingdom Lake (32° 53' N, 98° 30' W). Zoom in on the lake and fly around. Note the narrow lake within steep cliffs and the large number of docks and boathouses on the shoreline. To this point you have flown three hundred straight-line miles, while the tributaries and the mainstem of the river have meandered almost three times that distance.

Continue to follow the river along

its tortuous path through tree-covered hills and the rich old terraces and floodplains of the river, where people graze livestock, raise row crops, or plant huge pecan groves. You see the small towns of Mineral Wells and Palo Pinto, but they are obviously not river towns. The town of Brazos makes you think of a river town, but the bridge is closed and no roads lead to the Brazos River at Brazos, Texas.

Fly to Granbury, Texas, which is about 350 miles downstream from the headwaters on your direct course. Here you see Lake Granbury, long and narrow, passing through the city. As at Possum Kingdom, zoom in and note the large number of boathouses, canals, and docks.

Just downstream from DeCordova

Bend Dam, which forms Lake Granbury, you see the first major tributary join the Brazos from the west. This is the Paluxy River (32° 14' 48" N, 97° 43' 06" W). About 30 air miles downstream from DeCordova Bend Dam you reach the upper end (slackwater) of Lake Whitney (32° 04' N, 97° 29' W). The Brazos still looks like a river here rather than a lake. Fly around Lake Whitney, noting the steep limestone cliffs and the absence of structures on the shore. Your total straight-line distance since leaving Clovis, New Mexico, is about 390 miles. Most of the land now is in agriculture of various kinds. You still see few signs of people, but soon you will see another large lake, where the Bosque River is dammed at Waco just before it enters the Brazos in Cameron Park (31° 35' N, 97° 09' W).

The Brazos passes through the northeast side of Waco. A low dam backs the river up through the city, and there is a profusion of parks on both sides of the river. Waco is the largest city on the Brazos. At Waco the land changes drastically. Hilly and rugged with limestone cliffs, it suddenly becomes almost flat. Virtually all of the land is in fields and pastures except for the thick corridor of vegetation along the river. Downstream from Waco the river seems to wander aimlessly and you see many scars in the land where the river used to flow. However, zoom out and observe that the river forms a fairly regular, sinuous pattern.

Southeast of Waco you see Marlin, Texas, east of the river but not on it. Southwest of Marlin you can fly over the Falls of the Brazos (31° 15' N, 96° 55' W). About thirty straight-line miles downstream from the falls at Marlin you will see more falls at what was Port Sullivan, just west of Hearne, Texas (30° 52' N, 96° 42' W). The Little River, another important tributary, enters from the west about two miles downstream.

Slightly more than five hundred direct miles from Clovis, New Mexico, you pass Bryan–College Station, another large urban area, which is east of the river but not on it (30° 35' N, 96° 22' W). You see no parkland or any other public access to the river.

At 30° 23' N, 96° 10' W, you can see Hidalgo Falls, formed by a large limestone outcrop. Below Hidalgo Falls the Navasota River enters the Brazos from the east (30° 20' N, 96° 09' W). Observe how closely parallel the Navasota runs to the Brazos before they actually join. It won't be many years before a big flood changes the course of one of the rivers and they change their confluence. On the west side of the Brazos look at Washington-on-the-Brazos State Historic Site. Later you fly over Stephen F. Austin State Park (29° 49' N, 96° 06' W), just north of Interstate 10.

At this point the Brazos is a big, impressive looking river and amazingly "natural." There are no dams along the stretch from Waco to the Gulf, making it one of the longest reaches of undammed river in the United States. Although the surrounding countryside is all in agricultural use and fields or pastures often extend to the river's banks, there are many areas of floodplain forest adjacent to the river. There are also many meander scars and oxbow lakes. There seems to be little human interaction with the river. There are miles and miles where no roads approach the river. Brazos Bend State Park exemplifies these characteristics (29° 23' N, 95° 35' W).

Rather suddenly, at Richmond, you start to see roads and houses near the river (26° 36' N, 95° 46' W). The river continues its meandering, now again through agricultural land. It then sweeps smoothly by the old river towns of West Columbia and Brazoria. Fly to Freeport, Texas, and then zoom in (28° 57' 93" N, 95° 22' 27" W) to see where the Brazos was diverted past its old mouth to follow an almost straight course to the Gulf. Then fly around the old mouth, which is now Freeport Harbor. Note the canals that serve the petrochemical plants to the east of Freeport, as well as the Gulf Intracoastal Waterway, which intersects the Brazos near its mouth. You can also observe a large amount of standing water and some marshes, but there is no bay. The Brazos flows directly into the Gulf.

Your flight over the Brazos identified some important characteristics. Unlike many rivers, the Brazos comes from semiarid plains rather

It is possible to sit at Stephen F. Austin State Park and neither see nor hear anything other than the natural river environment, although the park is only forty-five miles from downtown Houston.

Water of the Brazos

If a river is water running downhill, we must know many things about the water to understand the river. Where does it come from? How much is there? When does it appear? What are the characteristics of its timing?

The Brazos receives only a small amount of its flow from groundwater, but that groundwater is important because most of it is very salty. The main source of water in the Brazos is rainfall. The characteristics of that rainfall and how it shapes the land are largely determined by the climate. Climate is the primary long-term factor that determines vegetation and soil characteristics as well as the land-shaping processes of erosion and deposition. Climatic effects are expressed locally, but they are the results of global processes and patterns.

ATMOSPHERIC CIRCULATION

The Brazos River is the product of rainfall in its watershed, but global processes determine the timing, amount, and often the absence of that rainfall. At the equator hot air rises up to an altitude of about ten miles, where the temperature may be minus 70 degrees Fahrenheit. The air cools as it rises, spreads toward the poles, and begins to descend at about 30° latitude, both south and north of the equator (Burroughs et al. 1996). The cooler, heavier air descends over the Brazos headwaters,

than snowy mountains or other major sources of water. The dams are in the upper half of the river. The lower half runs virtually free. There is no bay where it meets the Gulf, unlike some Texas rivers that do have bays. The most outstanding observation, perhaps, is how isolated the Brazos is from people for most of its course.

Weather Systems in Texas

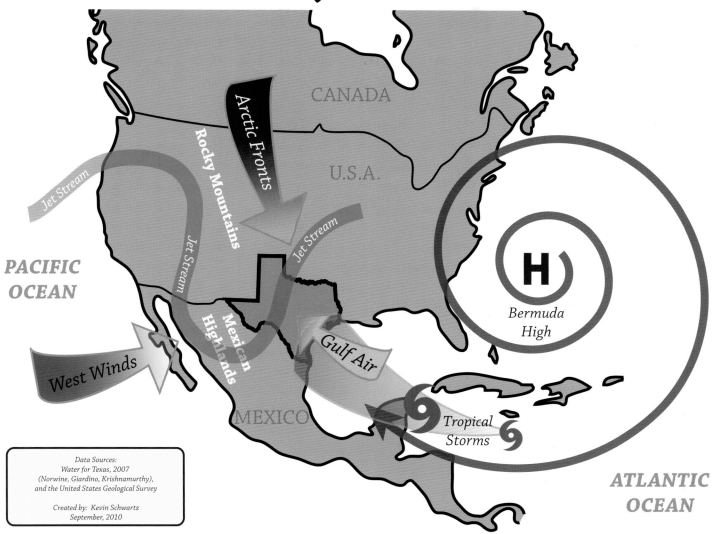

Data Sources:
Water for Texas, 2007
(Norwine, Giardino, Krishnamurthy),
and the United States Geological Survey

Created by: Kevin Schwartz
September, 2010

creating an atmospheric high-pressure area that prevents warmer moist air from moving in and thereby limiting the average annual rainfall in the region to about nineteen inches per year.

But it does rain nineteen inches at the headwaters on average and almost forty-eight inches at the mouth, so where does that rainwater come from?

The weather systems map shows several sources for that water plus the atmospheric circulation forces that move water into the region, overcome the high pressure, and initiate the processes that create rain.

While the upper Brazos basin is dominated by atmospheric high pressure, the lower half of the basin is dominated by air over the Gulf of

Mexico, which affects the land either by generating storms or simply in the conflict of its warmth and moisture with the cooler dry air from the north. But Pacific moisture comes in as well, as the jet streams pull air from the west. George W. Bomar, who literally wrote the book about Texas weather, summarized the processes: "The location of Texas with relation

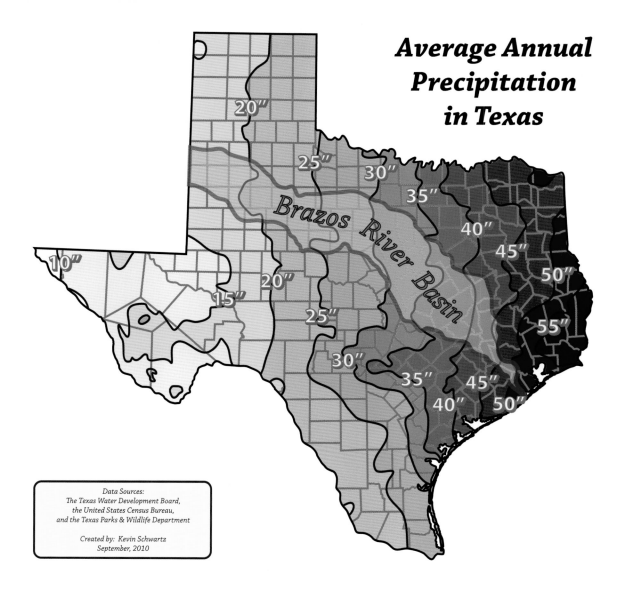

Average Annual Precipitation in Texas

Brazos River Basin

Data Sources:
The Texas Water Development Board,
the United States Census Bureau,
and the Texas Parks & Wildlife Department

Created by: Kevin Schwartz
September, 2010

to the North American continent, the warm Gulf of Mexico, and the not-far-distant Pacific Ocean guarantees a constant exchange of settled and unstable weather" (Bomar 2010, n.p.). Although quite variable during the short term, rainfall does form a general pattern over the long term. As shown on the precipitation map, the Brazos flows from dry to wet parts of Texas.

TEMPERATURE

Rainfall is the predominant weather characteristic that determines if the river runs, but temperature is important as well, especially as it relates to humidity and evaporation. For almost all of the Brazos basin the potential evaporation rate is higher than the rainfall. As described below, this fact has major implications for the vegetation and soils of the basin as well as the river, especially where it is dammed.

Water returns to the atmosphere from the earth's surface by evaporation and by release from plants (transpiration), a system known as evapotranspiration. A plant is a solar-powered pump that pulls water and nutrients from the soil to the leaves or other green parts, where they combine with carbon dioxide and light in

14

Chapter 1

the process of photosynthesis to produce the hydrocarbons that power much of the rest of life on earth in one form or another. It is necessary for the plant to release water so the water from its roots will move upward to the photosynthetic parts. A plant will release water as long as water is available in the soil. Where the potential evapotranspiration rate is higher than the available water, plants have several adaptive strategies. They can store food and survive through times when there is insufficient water in the soil to operate their life-giving pumps. They can minimize water loss, as do cacti, or they can be so effective in gathering water that other plants cannot compete, which is the strategy of junipers or creosote bush. Or, they can do like most people and live where they can get water. Some do this with deep roots, others by living close to a water source like the Brazos.

The relationships between climate, soil, vegetation, and hydrology are complex and not always obvious. For example, most of the biomass of grass is below ground, in the roots. Likewise, soil bacteria, algae, fungi, protozoa, nematodes, earthworms, and other wonderful creatures constitute most of the biodiversity of agricultural land.

Native plants in the Brazos basin, as everywhere, have had millions of years to evolve ways to respond to fluctuations in soil moisture. Still, the plants will die if extremely dry or wet conditions continue for too long,

but conditions change. Fire used to be common in much of the Brazos basin. Lightning set many fires, and Indians commonly used fire for hunting. Fire suppressed juniper and mesquite, but European settlers suppressed fire by overgrazing the grass fuel and by actively extinguishing the fires, so juniper and mesquite have greatly expanded their range across the Brazos basin. On the other hand,

native plants may be pushed out by newcomers, such as salt cedar from the Middle East, that are tougher and more aggressive and have colonized many of the stream courses of the Brazos drainage. Humans, of course, are the major agents of change in the

In the riparian zone the river's water supports vegetation, which consumes carbon dioxide to produce oxygen and provides a diverse forest habitat.

vegetation of the Brazos watershed and certainly qualify as an invasive species. Farmers have replaced almost all of the native grasslands of the basin with irrigated or nonirrigated row crops or pasture.

CLIMATE CHANGE

We are currently conscious of the probable effects of human-induced climate change, but we also recognize that such change is within the context of natural climatic cycles and anomalies. These changes take place over periods as short as decades to some as long as thousands of years.

A global climate change called the Altithermal occurred during the period from about 4,500 to 6,500 years before the present (BP). During that time much of the Brazos basin was very dry. After the Altithermal until about 2,000 years BP the climate became cooler and wetter than now and then returned to conditions similar to the present time (Holliday 1987; Carlson 2005). Climatologists consider the time between the late 1940s and 1957 as the "drought of record" for Texas. However, in the upper Brazos basin "for the past 6.5 years, streamflows in the area have averaged between 31 and 52 percent of flows occurring during the first 6.5 years of the previous drought of record" (Brazos G Regional Water Planning Group 2006). The year after this report, 2007, brought record rainfall, but 2008 saw a return to drought conditions and 2009 brought record

low flows in the lower Brazos. Are these short-term anomalies one of the periodic droughts of the region or the beginning of another event like the Altithermal? And, what might be the effects of human-induced atmospheric warming? Whatever the answers to these questions, the Brazos will continue to be governed by the climate.

Land of the Brazos

Most southwestern rivers, like the Pecos, Rio Grande, and western Colorado, run *through* the land, carrying water from high-elevation snowmelt. They receive relatively little water from surface runoff because much of the land they traverse is arid or semi-arid. Even those rivers, however, are strongly related to the land because they provide a unique aquatic habitat amid dry land and because people divert their water for irrigation and other uses.

The Brazos and its cousins the Texas Colorado and the Trinity run *off* the land as well as running through it. Their primary source of water is runoff from their watersheds or rather their catchment basins. Even the groundwater they may receive from spring flow is related to the land. Thus, to understand the Brazos we must understand the land that supplies it.

Water running off the land has a major role in shaping the land. The

flowing water of the Brazos has been eroding, transporting, and depositing earth material for ten million years. As we will see, these processes built much of Texas from the Balcones Escarpment to the Gulf of Mexico. As both a product and a producer of the land, the river has complex and often subtle relationships with the land, and they are more pronounced in the drier parts of the basin.

Next, we will set the context of the Brazos basin for the past 320 million years and summarize the relationships between the river and the natural regions through which it flows.

EARTH MOVEMENTS

The Brazos River basin sweeps in a smooth arc across the south-central part of what we now call the United States. It crosses six ecoregions, each with its specific geographic characteristics. But it lies within a broader context of geology and time that shapes its present and future characteristics.

The geologic characteristics of the Brazos basin are shaped by two kinds of earth movements. The first consists of the massive and extremely slow movements of the Earth's plates (tectonic movement). We cannot see those processes at work because they are so slow, but we can see their effects on the land. The southeastward slope of the land, which determines the river's path, is the result of geologic uplifts that formed the Rocky Mountains beginning about sixty-five million years ago.

The second type of earth movement is much faster and more localized. We see it when muddy water flows from a construction site or when riverside homes are washed from their foundations and stacked against one another. These processes of erosion, transport, and deposition of earth materials are called *fluvial geomorphology* and are the real "work" of rivers. Fluvial geomorphology refers to the processes by which flowing water shapes the land. We often don't like the results and take drastic efforts to prevent rivers from doing their work, but in the long run the river will do its job. As the river moves materials it continually reforms the river channel. In the process it creates, destroys, and re-creates a variety of habitats that provide opportunities for the great diversity of life forms and processes discussed in the next chapter.

Understanding the basic concepts of fluvial geomorphology helps us understand the behavior of rivers and the dynamic life processes of rivers, including our own relationship to them. The Brazos River is a particularly good example of fluvial processes for two reasons. First, it runs through a landscape that is easily eroded, and second, its frequent high flows have given it the power to shape the land.

Water's ability to move material depends on the speed of its flow. Generally the faster the water flows, the larger the size and quantity of materials it can move. A river has the greatest ability to move materials in times of bankful high flows, which may occur on an average of every 1.5 years (Wetzel 2001). But high flows are usually of short duration and may be relatively localized, so the moderate but more frequent flows may actually move most of the materials that are transported (Hudson and Mossa 1997). Thus, the river moves material some distance downstream and deposits it in a location where it will remain until it is moved again. The materials on the move may result from new erosion in the watershed, but more commonly they are materials that were deposited by previous high flows. These materials come from either the bed or the banks of the river.

The materials in the river's bed are especially subject to being moved because they are already saturated with water; a faster flow can thus more easily pick them up or push them along. Unseen to us, the river adjusts the capacity of its channel by scouring out materials as the flow increases but then depositing materials as the flow decreases.

Bank materials are also subject to the erosive power of water on the outside of a curve, where the flow is the fastest. It is common on the Brazos and its tributaries to see what we call *cut banks,* where the river is actively eroding the bank.

Since higher water velocity moves materials, it follows that a decrease in velocity will cause materials to be deposited. Anything that slows the river's flow will cause deposition. As the water flows around a bend the outside flow is faster, but the flow on the inside of the bend is slowed by friction with the bank. Thus, the river deposits material there, forming a *point bar.* Likewise, a tree trunk in the river will slow the flow and perhaps form a deposit called a *channel bar.*

As you flew along the Brazos in Google Earth® you noted the channel's sinuous meandering pattern, like a snake or a wave form. River scientists describe two common patterns of channels: braided and meandering. The upper parts of rivers, including those of the Brazos, may have relatively shallow and fast flow that forms unstable braided channels. However, meandering is the most common pattern for all rivers (Leopold 1994), and the Brazos is no exception.

The sinuous, meandering bends that rivers form are the result of complex physical interactions between the resistance of the river bank and the energy of the flowing water (Leopold 1994). Meandering channels are relatively stable but not permanent. The continual movement and deposition of material, especially at bends, causes the river channel to move as the outer bank of a curve is cut and material is deposited on the inside of the curve. This makes the curves ever tighter, so the river moves closer to itself. Ultimately, the neck between two curves can become quite narrow and may be cut by a large flood. In this case the curve is cut off,

The Brazos continues to cut its banks, a natural process that is a problem for farmers.

forming an oxbow lake. Fly Google Earth® to College Station, Texas. Just southwest of the city you will see two bends in the river that will probably be cut off in the not-too-distant future. Then fly down the river, where you will see oxbow lakes that have recently been cut off from the river. If you look closely you will also be able to identify old oxbow lakes that are now filled in but have darker soil and different vegetation. These are called *meander scars.* The oxbow lakes and even the meander scars provide important habitat diversity for the Brazos ecosystem.

Because rivers twist and turn, it is difficult to refer to a river bank by our usual directional terms of north, south, east, and west. Thus, we may use the terms left bank and right bank, with the left bank on our left as we head downstream. So, the Left Bank in Paris is a geographic term, not a political description.

DEEP TIME

Geologists estimate that the oldest rocks in the Brazos region are about 1.2 billion years old, from the Precambrian era (Hentz 2007). The oldest rocks actually visible in the Brazos basin are in the range of 286 million

to 320 million years old, from the Pennsylvanian period (Renfro 1979; Spearing 1991). Some of the limestone cliffs at Possum Kingdom Lake were deposited during the Pennsylvanian period. We will begin with the Pennsylvanian period, although almost 900 million years of geologic history preceded that time. The geologic history sidebar summarizes the processes and events that set the stage for the modern Brazos River basin.

Now let's look at the drama that has taken place on this geologic stage. Like all good plays this one is sensitive to its time and place. In this case the action is especially sensitive to climate. The physical geography sidebar summarizes the geographic characteristics of the Brazos basin.

Water Flow in the Brazos

Morton Salt® still uses its longtime slogan, "When it rains it pours." The Brazos is similar in that when it rains, it runs. The river receives relatively little spring flow, and what it gets is mostly salty. Thus, the precipitation that falls in the watershed is the primary source of the river.

Hydrologists measure water flow several ways, but a common measure is cubic feet per second (cfs). Think of the following to help you visualize flow. One cubic foot of water is about 7.5 gallons, so a flow of one cubic foot per second would be 450 gallons per minute, or about one full hot tub per minute. The maximum flow of the Brazos ever recorded was 246,000 cfs, the equivalent of almost a quarter million hot tubs per minute.

The average flow actually tells us little. The peak and low flows are the ones that determine many of the important characteristics of a river, including how we humans may relate to the river. Flows greater than ten

The energy of flowing water continually reshapes the river's channel and its banks. Sinuous curves called meanders are the most common channel form.

Years before present (geologic period)	Events and processes
286–320 million (Pennsylvanian)	Collision of plates uplifted the Ouachita Mountains along a line that would connect Fort Worth, Austin, San Antonio, Uvalde, and Marathon. To the west of these mountains a succession of seas covered the area, depositing limestones, shales, and sandstones.
245–286 million (Permian)	The land continued to rise, constricting the seas west of the mountains and filling them with sediment from erosion. Shales, salt, and gypsum were deposited in the area that would later become the upper Brazos drainage.
208–245 million (Triassic)	Shales and sandstones were deposited in what is now the Llano Estacado.
144–208 million (Jurassic)	There are no Jurassic rocks in the Brazos basin. The Gulf was a shallow sea that repeatedly dried, depositing salt that later extruded upward, becoming the salt domes of the Gulf Coast.
65–144 million (Cretaceous)	Tectonic movement pulled the continents apart. A succession of seas covered the region during this period. The last one, the Bearpaw Sea, reached from today's Gulf of Mexico to the Arctic Ocean. Massive layers of limestone were deposited. Geologic evidence indicates that about 65 million years ago an asteroid 6 miles in diameter, moving at 56,000 mph, came from the southeast and hit the shallow Bearpaw Sea near what is now Chicxulub, on the Yucatan Peninsula. Debris and heat from the impact destroyed most life in North America and ended the age of the dinosaurs, but much of the aquatic life survived and continues to live in the modern Brazos. Tsunamis resulting from the impact caused unusual geologic depositions, marking this K-T boundary (Cretaceous-Tertiary). Geologists found examples of these deposits near the Brazos in Falls County.

2–65 million (Tertiary)	Flowing water from the northwest deposited massive amounts of sand and mud in the Gulf, building the land from just southeast of Marlin to Freeport.
	
	    Erosion from new uplifting of the Rocky Mountains deposited large amounts of sediment on the High Plains, which resulted in the Ogallala formation.
	
	    About 10 million years ago the Brazos, the Colorado, and the Red rivers began to carve their valleys. They cut away at the High Plains, moving the Caprock Escarpment about 200 miles west to its current location.

0–2 million (Quaternary)	In the period of approximately 125,000 to 10,000 years ago (Late Pleistocene) the flow of the Brazos may have periodically been four times the current flow, causing it to erode a larger basin than would be eroded by the modern river.
	
	    The mouth of the Brazos was realigned in 1929, isolating the original mouth as Freeport Harbor; the new mouth is 6 miles southwest down the coast.
	
	    The Brazos carries the largest sediment load of any river in Texas. It filled its estuary and now empties directly into the Gulf. It has extended the shoreline about 1.5 miles at the new mouth.

Sources: Renfro 1979; Bourgeois et al. 1988; Spearing 1991; Dunn and Raines 2001; Flannery 2001.

Although not the oldest rock in the Brazos basin, Cretaceous limestone forms beautiful cliffs at Lake Whitney.

times the average occurred at Waco seventy-five times between 1899 and 2007. The peak flow of the Brazos (the 246,000 cfs record) occurred at Waco on September 27, 1936, and the lowest monthly average flow there was 33.3 cfs in September 1999 (U.S. Geological Survey 2008). Thus, the one-time high flow was 7,387 times greater than the lowest average flow at Waco.

The flow data tell us that the dominant characteristic of the Brazos River is frequent flooding. Especially

Location	Ecoregion and its characteristics	Dominant vegetation
Eastern New Mexico to the Caprock Escarpment, approx. 130 direct miles	High Plains: Almost level plateau, elevation from 4,500 to 3,000 feet above sea level, sloping gently to the southeast Numerous playa lakes Average annual rainfall 15–21 inches, with times of extended droughts	Irrigated crops and rangeland
Caprock Escarpment to just above Possum Kingdom Lake, approx. 180 direct miles	Rolling Plains: Gently rolling to moderately rough topography cut by narrow stream valleys running east and southeast Elevation 3,000 to 800 feet Average annual rainfall 22–30 inches	Mesquite-lotebush, with yucca, sumac, juniper, bluestems, and grama; also three-awn grasses
Above Possum Kingdom Lake to Waco, approx. 170 direct miles	Cross Timbers and Prairies: Gently rolling to hilly Elevations 800 to 400 feet Average annual rainfall 24–34 inches	Post oak, juniper, mesquite, and live oak in woodlands Silver bluestem and Texas wintergrass on the prairies
Waco to approximately 20 direct miles below Marlin, approx. 40 direct miles	Blackland Prairie: Gently rolling to nearly level topography Elevations 400 to 350 feet Average annual rainfall 30–40 inches	Virtually all of the Blackland Prairie is now agricultural land, with row crops and improved pasture.
Northeastern Milam County, near Cameron, to about 20 direct miles below Bryan–College Station, approx. 55 direct miles	Post Oak Savannah: Gently rolling to nearly level topography Elevations from 350 to 200 feet Average annual rainfall 40–50 inches	Highly diverse woodland and grassland mosaic, including post oak, blackjack oak, hickory, eastern red cedar, cedar elm, and bluestem, lovegrass, panicum, and three-awn grasses
Southeastern Brazos County to about 15 direct miles below Hempstead, approximately 40 direct miles	Blackland Prairie: Gently rolling to nearly level topography Elevations from 200 to 100 feet Average annual rainfall 40–50 inches	Native and introduced grasses
From below Hempstead to the mouth, 80 direct miles	Gulf Prairies and Marshes: Nearly level, slowly drained plain Elevation less than 150 feet Average annual rainfall up to 50 inches	Cropland on uplands Pecan-elm forest in river corridor

Sources: Adapted from McMahan, Frye, Brown 1984; http://www.tpwd.state.tx.us/nature/tx-eco95.htm

The waters of the Brazos flow from a variety of landscapes as the river traverses Texas.

large floods occurred in 1833, 1842, 1913, 1921, 1991, 1995, 1998, and 2007. This flooding is the result of the climate and the land, discussed earlier. Flooding was the major factor in determining our historic relationship with the river, as we will explore later.

DAMS AND RESERVOIRS

As we will see in chapter 5, the flow of much of the Brazos is regulated by dams. The Brazos has more dams than any other river in Texas. People have been building dams on rivers for almost five thousand years to the best

FLOW FACTS

Average flow near river's mouth

Brazos	7,671 cfs
Mississippi-Missouri	619,339 cfs
Rio Grande	2,895 cfs
Colorado (western)	5,930 cfs

Brazos River flow variations

Lowest annual average near the mouth	990 cfs (2000)
Highest annual average near the mouth	29,050 cfs (1992)
Peak flow near the mouth	123,000 cfs (June 6, 1929)
Peak flow at Waco	246,000 cfs (September 27, 1936)
Peak flow of Clear Fork	149,000 cfs (August 4, 1978)
Peak flow of Double Mountain Fork	91,400 cfs (September 26, 1955)
Peak flow of Salt Fork	35,600 cfs (May 24, 1954)
Peak flow of Running Water Draw	12,000 cfs (June 6, 1941)

of our knowledge. A dam with the Arabic name Sadd el-Kafara was built south of Cairo sometime between 4,760 and 4,960 BP (Smith 1971). We build dams because the natural flow of rivers often does not meet our needs or desires. The water is not where we want it, when we want it. The earliest dams were built to retain water for irrigation and other uses. Later, people began to build dams to generate power, using equipment ranging from simple water wheels to today's sophisticated turbines. Dams were built to raise the water level over shoals to facilitate navigation. More recently, in the twentieth century, we built dams to detain water and thus prevent downstream flooding. The dams on the Brazos represent all of these purposes; they were often constructed as multipurpose projects to generate electric power, detain flood-

waters, and retain water for various uses. The terms *detain* and *retain* have specific meanings when applied to dams. Floodwaters are detained until the flood recedes and the water can be released. Stored water is retained until it is needed for its specified use.

The water held back by a human-constructed dam is properly called a reservoir, not a lake. A lake is a body of water that forms in a natural topographic depression or behind a natural dam such as a logjam, beaver dam, or rock slide. Lakes usually receive their water from surface runoff or groundwater and are not as directly affected by river flow as are reservoirs. In fact, it is common for a natural lake to create a river with its overflow, such as Lake Itasca, which generates the Mississippi River. Lake levels usually do not fluctuate as often or as greatly as

those of reservoirs. Lakes are usually clearer than reservoirs because their inflow water contains fewer sediments, but they are also usually less productive than reservoirs for that same reason. The only natural lake in Texas is Caddo Lake, which was formed originally by a logjam, although people stabilized the dam in the 1940s. However, regardless of technicalities, in Texas we call our reservoirs "lakes." I use both terms throughout this book. If I need to emphasize that the impoundment is not natural I use the word *reservoir*, but the rest of the time I call the bodies of water "lakes."

Dams and their reservoirs are designed and managed to achieve specific purposes. Reservoirs are typically divided into horizontal *pools*, each of which is dedicated to a specific purpose. The pools are not enclosed like a swimming pool but are defined by water elevation and capacity. Generally there are three pools from the lowest to the highest elevation: power pool, conservation pool, and flood pool.

The *power pool* is the water that is allocated to be released to generate hydroelectric power according to whatever agreements the dam operators have regarding power production.

The reservoirs (lakes) on the Brazos were built primarily to store water and reduce floods. As seen here at Possum Kingdom Lake, their levels vary according to rainfall and the demands for water by downstream users.

This can mean that downstream flow is determined by the demands for power, not by ecological or recreation needs.

The *conservation pool* holds water for municipal and industrial use, irrigation, and recreation. The reservoir is managed to try to keep this pool as full as possible, but most of the water in the pool is owned by downstream rights holders. When they request their water, it must be released. In recent decades the conservation pools of most reservoirs have become important for recreation. Recreation interests want the water to be retained in the reservoir, and conflicts arise about how much water should be released and when. This will become an increasingly important issue as the state develops its policies to maintain instream flow in the rivers.

The *flood pool* is the upper pool and is left empty so it can detain water and reduce the downstream effects of floods. This means that the land within the flood pool is usually dry but will occasionally be flooded. Recreational facilities such as campgrounds are often built in the flood pool, but structures that would be damaged by floods are not (or should not) be built there. Lake Whitney has a large flood pool, and thus there are very few structures near the water.

Water Quality

The Brazos is muddy and salty but not nasty, as some people think it is. Even an employee at a state park on the Brazos advised us to go to the Colorado River if we wanted to experience a "nice" river. The water is not pristine by any means, and some parts of the river do not meet the state standards for "contact sports," that is, swimming. However, the Brazos is not too bad, considering that it receives the effluent from several urban areas and runoff from forty-two thousand square miles, virtually the same land area as Tennessee.

Natural rivers have an amazing ability to cleanse themselves. Biological, chemical, and physical processes cause many pollutants to be taken up by aquatic plants or changed chemically. These processes can function indefinitely, but their capacity can be overwhelmed if excessive pollutants are put into the river. The ultimate cleansing is when the pollution input ceases and the river truly renews itself. Thus, even though the pollutants in the Brazos exceed state standards in some areas, those problems are solvable.

However, we probably cannot eliminate the river's mud, and so far we have not been able to reduce its salt. The salt is natural and therefore is not really a pollutant. The mud is mostly natural, but the intensive row cropping in much of the basin exposes soil to erosion. For example, the Clean Rivers Report shows a median value for total suspended solids at Waco of 18 milligrams per liter (mg/L) (Brazos River Authority 2007). Waco is only a few river miles below the lowest of three major dams on the mainstem of the Brazos, dams that trap the sediment from the upper river. But 179 river miles downstream, as the river and its tributaries flow through rich agricultural land, the silt load increases three and a half times, to a median value of 66 mg/L for total suspended solids. This muddiness of the river is technically called turbidity and results primarily from the presence of extremely small particles of soil that are suspended in the water. The size and amount of these suspended particles are partially determined by the speed and turbulence of the flow. Microscopic plants in the water (phytoplankton) also produce turbidity.

Having reviewed the Brazos River in its atmospheric and geologic contexts, we now turn to the life processes of the Brazos within the context of an integrated ecological system.

CHAPTER 2

The Brazos as an Ecological System

Ecosystem: A community of organisms together with their physical environment, viewed as a system of interacting and interdependent relationships and including such processes as the flow of energy through trophic levels and the cycling of chemical elements and compounds through living and nonliving components of the system.

American Heritage Science Dictionary

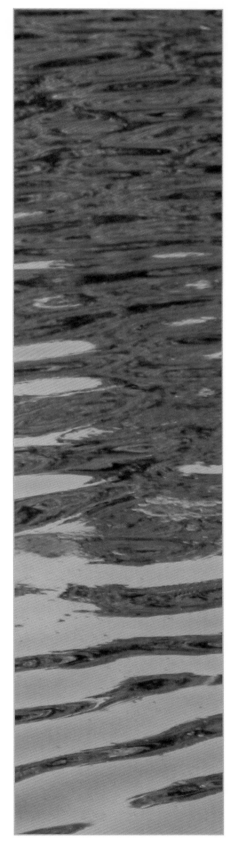

ALTHOUGH THE BRAZOS River crosses six of Texas' ecological regions, we can also think of the river and its watershed as an integrated ecological system (ecosystem). In most cases the boundaries of an ecosystem are rather indistinct. River ecosystems, however, are defined by the boundaries of their watershed or catchment area. Because water runs downhill to the river, everything within the watershed is to some degree part of the river's ecosystem.

A river's ecosystem is defined by the relationships between water and the characteristics of the watershed. The water we see flowing downstream is not the only water in the river's ecosystem, perhaps not even the most important. Water saturates the bed and bank materials and sometimes goes over the banks into the floodplain and then returns to the river. The water varies in sequence, volume, velocity, duration, direction, temperature, clarity, oxygen content, and chemical characteristics. The water in the river's ecosystem relates with seasons, sunlight, temperature, wind, rocks and soil, vegetation, and human modifications of the watershed. All of this creates a grand and dynamic complexity of opportunities and challenges for life in the river, resulting in an amazing diversity of life forms and processes.

Rivers are probably the most dynamic ecosystems on Earth due to their constant variability. The water is continuously moving downstream, seldom flowing straight, often moving in complex back currents, and even spiraling down the channel like a corkscrew or spinning around in whirlpools and eddies. The flow may actually stop at times, and the river habitats will become dry and sunbaked. But later the river will flood, scouring out habitats and then depositing new sediment. The depth will vary with flow and with the characteristics of the channel. The water will vary between muddy or relatively clear, cold or warm. The chemical characteristics will vary daily and seasonally. The depth, width, and path often vary with an almost mathematical rhythm.

This huge variability requires that river organisms be resilient and tolerant of the changing conditions of their habitat. It also presents opportunities for specialization. There are fish for the deep pools and the shallow riffles. There are insects that walk on water and others that hide under rocks. There are microscopic life forms that live between the minute particles of bottom sediment and others that stick to rocks. Green plants produce their own food by photosynthesis, other

An egret watches for potential prey navigating the Falls of the Brazos near Marlin.

things eat the plants, and yet others eat the animals that eat the plants. Some live by eating the organic trash washed into the river or the wastes of others. In North America there are about 10,000 freshwater aquatic insect species, 1,000 native fish species, about 340 types of freshwater snails, almost 340 varieties of crayfish, and almost 300 species of freshwater mussels (Benke and Cushing 2005).

This chapter will help readers gain an understanding and appreciation for the ecosystem processes of the Brazos River. We begin with general concepts and work our way down to a certain level of detail—not too deep because ecosystems are immensely complex.

General Concepts

Freshwater ecologists (limnologists) use three concepts to help explain the flows of materials and energy within a river ecosystem: (1) the river continuum concept, which describes the movement of materials and energy down the river, (2) the flood pulse concept, which describes the movement of materials and energy laterally from the floodplain, and (3) local riparian production, that is, the biological production of materials and energy within the river and the riparian zone (the area adjacent to the river that is directly affected by soil moisture from the river). These various energy pathways depend partly

on fluvial geomorphology and partly on the characteristics of the riparian corridor and the floodplain.

RIVER CONTINUUM CONCEPT
The river continuum concept recognizes that upper, middle, and lower parts of a river (reaches) have different physical and biological characteristics and that upstream characteristics partly determine certain downstream characteristics (Vannote et al. 1980). The upper reaches of a river are often narrow and shaded and thus have relatively little primary production from photosynthetic algae and plants. Biological productivity depends mostly on organic matter washed into the river. This organic matter may range in size from tree trunks to particles one millimeter in diameter. The species in the upper reaches have evolved in order to use this coarse particle organic matter (CPOM). An aquatic system that depends on organic matter from outside sources is called *allochthonous*.

The middle reaches of a river are generally wider and more exposed to sunlight. They may be relatively shallow and somewhat clear, so light can penetrate the water and support primary production via photosynthesis, which provides most of the energy for life in these sections of the river. The species that make up the river community reflect this local energy base. An aquatic ecosystem that produces its organic matter within the system is called *autochthonous*.

Much of the Brazos's floodplain in Waco is parkland, which is intended to maintain most of the river's natural functions and to minimize damage to structures and facilities.

The upper, middle, and lower parts of a river have different ecological characteristics and functions. These natural characteristics have been modified by dams in the central part of the Brazos.

In the lower reaches a river is usually deeper than it is in the upper reaches. The lower reaches also carry silt brought in by the upper river and all of the tributaries, including the waste from the species that used the coarse particles, which has been converted to fine particle organic matter (FPOM) that consists of pieces less than one millimeter in diameter. The turbid water prevents light from penetrating very deeply, so photosynthesis is limited and the species rely on the fine particle organic

matter as the basis of energy, so the lower river reaches are *allochthonous*, like the upper reaches.

FLOOD PULSE CONCEPT

While the river continuum concept emphasizes the continuous dynamics of downstream flow, the flood pulse concept recognizes the contribution of organic matter carried into the river from the floodplain as periodic floods recede (Benke et al. 2000). A natural floodplain is a rich storehouse of fine soils and a variety of organic matter that decomposes between floods and supports new growth as well. Some of these materials are picked up by a flood and brought into the river's flow. At the same time, the river deposits new materials on the floodplain.

LOCAL BIOLOGICAL PRODUCTION

Organic matter flows down the river and enters from the floodplain, but a substantial amount also comes from photosynthesis in the water and especially from riparian vegetation (Thorp and Delong 1994). Even in the uppermost reaches of the Brazos, where rainfall is relatively low, streamside vegetation is lush because the soil absorbs water from the river. Leaves, seeds, and bird and insect droppings fall directly into the river. Quite often entire trees or large limbs fall into the river and are lodged against the bottom and the bank. Limnologists have

This 1917 scene looks across the same floodplain area as in the modern photograph and illustrates the abrupt geologic and topographic division marked by the Brazos River through Waco. Waco's famous Lovers' Leap cliff overlooks the Bosque River, which enters the Brazos at the end of the line of trees.

recently recognized that large woody debris provides important habitats; it also ultimately decomposes and contributes organic matter. These large items work like sieves to trap other debris and sometimes canoeists, who politely call them "strainers" but more often use stronger words for these dangerous river features.

THE CRUCIAL IMPORTANCE OF FLOW VARIATIONS

Life in the Brazos has evolved over millions of years in response to the climate of its drainage basin. As de-scribed in the previous chapter, the climate has changed during that time, but the river adjusts its geomorphological characteristics rather quickly, at least in geologic time. In addition, ecological processes evolve to take advantage of the flow and geomorphic characteristics. As described below, river scientists now recognize the ecological importance of short-term natural variations of flow (Postel and Richter 2003).

In their analysis of river flows that are essential for maintaining a sound ecological environment in the Brazos and other Texas rivers, a consortium of state agencies described four levels of flow, the basic functions of each, and the timing, as outlined in the table on flow level processes and consequences.

For at least several thousand years rainfall in what is now Texas has been concentrated in the spring, with a smaller rainy season during hurricane season in late summer and early fall. Texas river ecosystems have evolved to be most productive, and reproductive, in the spring, when water is most abundant. Subsistence flows are the lowest flows that will prevent excessive concentrations of pollutants and will provide sufficient water to maintain aquatic habitats. Texas rivers can tolerate subsistence flows for the last half of the year.

Base flows are those that should occur in January through May, providing dependable conditions for production and reproduction processes. Many river species depend on certain velocities of flow for feeding. Also, water temperature and dissolved oxygen content are closely related to flow velocity. Aquatic species may have specific requirements for temperature and dissolved oxygen. Because the river is the primary source for water that supports the riparian community, it is essential that a certain level of flow be maintained so that the bank can absorb water from the river.

High flow pulses are major determinants of river characteristics. Typically, they occur in the spring, and life forms have evolved to take advantage of the higher flows. Also, high flow

Vegetation at the river's edge, as seen here near Waco, produces an important part of the river's organic matter and provides microhabitats for a variety of species.

Flow level	Processes and consequences of flow levels
Subsistence flows	**Hydrology:** cause infrequent, low flows **Geomorphology:** increase deposition of fine organic particles **Biology:** provide restricted aquatic habitat; limit connectivity **Water quality:** elevate temperature and constituent concentrations; maintain adequate levels of dissolved oxygen
Base flows	**Hydrology:** create average flow conditions, including variability **Geomorphology:** maintain soil moisture and groundwater table **Biology:** provide suitable aquatic habitat; provide connectivity along channel corridor **Water quality:** provide suitable in-channel water quality
High flow pulses	**Hydrology:** cause in-channel, short duration, high flows **Geomorphology:** maintain channel and substrate characteristics; prevent encroachment of riparian vegetation **Biology:** serve as recruitment events for organisms; provide connectivity to near-channel water bodies **Water quality:** restore in-channel water quality after prolonged low flow periods
Overbank flows	**Hydrology:** cause infrequent, high flows that exceed the channel depth **Geomorphology:** provide lateral channel movement and floodplain maintenance; recharge floodplain water table; form new habitats; flush organic material into channel; deposit nutrients in floodplain **Biology:** provide new life phase cues for organisms; maintain diversity of riparian vegetation; provide conditions for seedling development; provide connectivity to floodplain **Water quality:** restore water quality in floodplain water bodies

Source: Texas Commission on Environmental Quality, Texas Parks and Wildlife Department, and Texas Water Development Board 2008.

pulses are the major shapers of the river channel due to their power to erode bed and bank material, transport it, and deposit it downstream. The physical shape of the channel, and consequently the opportunities for habitats, are determined by high flow pulses. Of course, sediments contain the nutrients that are essential to the river's ecological processes.

Overbank flows are necessary in order to maintain the viability of the floodplain community, which consists of species that have evolved to depend on and to tolerate flooding. We now know that the biological production of the floodplain is an important part of the river's ecological processes.

River Habitats

Stand beside the Brazos where it is free flowing and you will see ripples, eddies, perhaps riffles and deeper pools, and you will see the water flowing around and among a variety of vegetation in the shallows and along the river bank as well as snags in the river. Each of these provides different conditions of oxygen, current, exposure to light, water levels, and even nutrient availability.

Things we cannot see compound the complexity of the river ecosystem. Water saturates the fine sediments of the bed and bank as deeply as three feet or more in what is called the *hyporheic zone* (Wetzel 2001). The water

Each rock, riffle, gravel bar, and side pool provides constantly changing habitats for specialized plants and animals, as seen here at the Nichols crossing near Brazos Point.

The floodplain is also a riverine habitat but receives less frequent watering than the riparian zone. Generally, the water will not remain there for long, but in 1842 the lower Brazos remained out of banks for several weeks (Hendrickson 1981). On the Brazos the floodplain soils are fine and porous and absorb water rapidly and deeply. Thus, the floodplain habitat offers more water than the surrounding uplands, but the residents must be adapted to dry periods in addition to their times of abundant water.

The alluvial sediments of the Brazos River corridor absorb and hold a large amount of water, forming the Brazos River Aquifer. This aquifer extends about 350 miles, from southern Hill and Bosque counties to Fort Bend County and is as much as 7 miles wide and 85 feet deep. Wells drilled into the aquifer supply water for irrigation, with some producing as much as one thousand gallons per minute (Texas Water Development Board 2010).

The lower Brazos has meandered across a broad plain for most of one million years. During these meanders it has cut off bends and formed oxbow lakes, which ultimately fill in and form meander scars, which you can see from Google Earth®. Both the

and nutrients available in the mud support large populations of bacteria and other microscopic life forms. Some of these areas are constantly covered and supplied with water, but others dry out when the water is low, only to be covered again when the level rises.

The riparian zone parallels the river and thrives on moisture that is absorbed by the river banks. This absorption may result only in soil moisture but could also lead to the formation of an alluvial aquifer that can be tapped by long-rooted vegetation on

The flowing river slows as it enters the slackwater at the upper end of Lake Whitney and drops its sediment load, as shown by the stair-step sand deposit at the lower left of the photo.

oxbow lakes and the meander scars are river habitats because they provide soil and water conditions similar to the riparian corridor and reconnect with the river during high floods.

The modern Brazos and its tributaries have been dammed in many locations. The resulting impoundments range from stock ponds (earthen tanks) as small as an acre or less to the river's largest reservoir, Lake Whitney, with 23,220 acres in surface area. These bodies of water provide a variety of habitats quite different from the flowing river. The upper part of the reservoir, where the river first slows as it encounters the impounded water, is called the head of slackwater. This is where the river drops much of its sediment load and forms an unstable habitat that may be changed with each flood or may be exposed as the reservoir level drops. Possum Kingdom Lake and Lake Whitney were built in narrow, steep-sided valleys (almost canyons), so there are areas of deepwater habitat with relatively little shallow water along the shore. Where the land slopes more gently the impounded water forms shallow areas called *littoral*. Both the deep and shallow areas provide a variety of habitats. Also, coves where tributary creeks enter the reservoir can provide habitats that differ from the main body of the reservoir because organic matter is brought in by the creeks. The section of the Brazos that has been the most highly modified by dams is discussed in chapter 5.

Ecological Functions

When it comes to river life, probably we think about fish first because they are fun to catch and good to eat. Then perhaps we think about snakes because some of them may bite us. In the lower Brazos we might think about alligators because they can take a much bigger bite and are magnificent to see. But these obvious biological actors are neither the most abundant nor the most important to the resilience of the river ecosystem. The resilience of a river refers to its ability to continue to function biologically even during periods of substantial change such as a drought or flood. Rivers are resilient systems due largely to the variety of habitats described above and the diversity of life forms that have evolved to exploit those habitats. However, a river's resilience depends on the proper functioning of its natural processes and thus is limited. Our abuse and mismanagement of a river can diminish its resilience until it is dead—just a ditch of polluted water.

As in most systems, those at the bottom of the food chain are essential because they provide the energy for the system. Those species that either capture energy via photosynthesis or convert the organic material (detritus) washed into the river into usable energy are the fundamental bases of life in the Brazos. The bacteria living in the saturated sediments of the hyporheic zone, especially at the center of the river, play a major role in carbon and nitrogen cycling. This role is so important that researchers have called these bacteria the "river's liver" (Fischer et al. 2005). Photosynthetic life forms in the water itself include diatoms, unicellular green algae, filamentous green algae, and blue-green algae (Benke and Cushing 2005). We also know that all of the green plants growing in the riparian corridor, from the smallest mosses to giant cottonwood trees, participate in the primary production of the river.

The most common life forms in the Brazos are the macroinvertebrates, including insects and crustaceans that are usually larger than one-half of a millimeter (Benke and Cushing 2005). Macroinvertebrates comprise 96 percent of animal species on Earth. Many of them spend part or all of their lives in fresh water, from the mosquitoes that feed on us to the crayfish that we feed upon (Voshell 2002).

The job of many macroinvertebrates in the river is to convert detritus into matter that some other form of life can convert to energy. These *detritivores* are divided into shredders, filtering collectors, and gathering collectors (Benke and Cushing 2005). Shredders break down coarse particle organic matter into fine particle organic matter, which the filtering and gathering collectors then consume. Shredders in the Brazos include stoneflies, caddisflies, true flies (order Diptera, including such species as blackflies), and crustaceans.

Filtering collectors include mayflies, caddisflies, true flies, and mussels. Gathering collectors are predominantly insects of many types.

Some macroinvertebrates are grazers and others are predators. Grazers (also called scrapers) feed on the photosynthetic diatoms and algae growing on solid surfaces of rocks and large woody material. Grazers include caddisflies, mayflies, and snails. Predators kill and eat other animals. Macroinvertebrate predators include dragonflies, damselflies, stoneflies, hellgrammites (dobsonfly larvae), and true bugs (order Hemiptera, such as water striders and water boatmen).

All of the other animals associated with a river have similar functions. Some fish are predators, others grazers, and others filter feeders. Snakes and most mammals and birds are predators, except beavers, which eat bark and twigs. Alligators and turtles are omnivores that will eat almost any organic matter.

This "functional" description of river life may seem unusual. We more commonly describe biology by naming things (taxonomy). Lists of major species in the Brazos are provided in the appendix. However, we have a better idea of the river as a living system if we understand the ecological functions associated with it and the variety of ways those functions can be performed. Breaking aquatic life into the functional groups of shredders, grazers, and so forth seems confusing, especially when we observe that a group such as caddisflies fits in several categories. We are used to more specific job descriptions: cows are grazers and lions are predators. But we have to realize the extraordinary variety in river life, especially among the macroinvertebrates. For example, there are about fourteen hundred species of caddisflies in North America (Voshell 2002). There are at least forty-two species in the middle Brazos alone (Benke and Cushing 2005). Insects are highly specialized, so it should be no surprise that they have different jobs.

Learning Ecology from the Brazos

Any river is a good laboratory for learning ecology, but the Brazos is especially valuable because it is so diverse, providing a wide variety of learning opportunities. There are two obvious questions about learning ecology from the Brazos. First, what can we learn, and second, why might we want to know those things?

What can we learn about ecology from the Brazos?

- names and lifestyles of a wide variety of aquatic life, from alligators to zooplankton
- relationships among climate, geology, biology, and human activity
- basic water science, including flow, turbidity, temperature, and some chemistry
- how to protect the ecological health of a stream

What is the value of learning river ecology?

- deepens our understanding and appreciation of life processes
- can help people develop an interest in science, which may provide direction for students as they pursue their educations
- can lead to life-long interests such as fishing, birding, and insect and plant study

ACTIVITIES
Much basic river science involves observation and simple collecting activities that can be done by almost anyone. The resources listed below provide information and directions for a large number of activities using equipment that you can make at home. Most of these activities involve actually getting into the river or going out on it in a boat. However, it is quite possible to learn many things about the Brazos without getting in or on the water. Virtually any place where you can view the river will provide endless hours of learning about riparian vegetation, birds, insects, snakes,

With its constant variations in space and time a river offers a wide variety of habitats and supports a large number of species.

turtles, fish, frogs, and perhaps even alligators.

Many of the activities require shallow water that is easily accessible. Much of the Brazos upstream of Waco, between the lakes, is suitable. Shallow coves where creeks enter the lakes are good study sites, as are tributaries like the Paluxy and Bosque. Other learning activities can be done by simply observing the river or walking in the floodplains in places such as Stephen F. Austin and Brazos Bend state parks.

PLACES TO LEARN ABOUT THE BRAZOS

The Brazos and its tributaries provide an astounding number of sites that are suitable for learning about aquatic ecosystems and the region's history. Most of the places are on the lakes, but some are on the rivers.

Corps of Engineers

On its ten lakes in the Brazos basin the U.S. Army Corps of Engineers provides sixty-four separate areas with water access and hundreds of campsites. The Corps of Engineers provides detailed information about its recreation facilities on the Internet. Search for "Corps Lakes," a new site and access system recently put online.

The Brazos offers a variety of recreational opportunities that can be educational as well.

The Corps of Engineers lakes that offer recreation areas in the Brazos basin are:

- Lake Aquilla: 3 areas
- Lake Belton: 14 areas
- Lake Georgetown: 5 areas
- Granger Lake: 6 areas
- Navarro Mills Lake: 5 areas
- Proctor Lake: 4 areas
- Lake Somerville : 3 areas
- Stillhouse Hollow Lake: 7 areas
- Lake Waco: 6 areas
- Lake Whitney: 13 areas

The Corps of Engineers also maintains a wildlife management area at Lake Aquilla.

Texas Parks and Wildlife Department

The Texas Parks and Wildlife Department provides fourteen facilities that help educate people about life and history in the Brazos basin. Search for "Texas State Parks find a park" and select http://www.tpwd.state.tx.us/spdest/findadest/. There you can select regions of the state and specific information about individual parks. Also search for "Sea Center Texas" and select http://www.tpwd.state.tx.us/spdest/visitorcenters/seacenter/. State recreation/education facilities in the Brazos basin are:

- Abilene State Park
- Lake Mineral Wells State Park
- Possum Kingdom State Park
- Dinosaur Valley State Park
- Cleburne State Park
- Meridian State Park
- Lake Whitney State Park
- Mother Neff State Park
- Fort Parker State Park
- Lake Somerville State Park
- Washington-on-the-Brazos State Historic Site

Because the Brazos is a wild place, some reasonable precautions are necessary.

- Some of the banks are steep and unstable.
- The current may be strong.
- Poison ivy is often abundant.
- Snakes will bite if cornered. Cottonmouth water moccasins can be aggressive.
- Most of the Brazos runs through private property. Do not enter private property without permission from the landowner. Purple paint on signs, fence posts, or gates means the property is "posted" and you must not enter.
- Bridges appear to be good places to view the river, but traffic can be a hazard.

- Stephen F. Austin State Park
- Brazos Bend State Park
- Sea Center Texas

The Texas Parks and Wildlife Department offers information on the Internet about lake fishing, and the site also provides other useful recreation information. Search for "Texas Lake Finder" and select http://www.tpwd.state.tx.us/fishboat/fish/recreational/lakes/. There you can click on regions of the state and individual lakes. The site lists characteristics of each lake and a link for public access.

The Texas Parks and Wildlife Department operates the Granger Lake Wildlife Management Area (10,888 acres), Lake Somerville Wildlife Management Area (3,180 acres), and the Justin Hurst Wildlife Management Area (11,938 acres) near Freeport. Get information on those wildlife management areas by searching for "Texas find a wildlife management area" and selecting http://www.tpwd.state.tx.us/huntwild/hunt/wma/find_a_wma/#map

Texas Historical Commission
The Varner-Hogg Plantation is located at West Columbia.

Wetlands are important parts of the river environment that are misunderstood and have frequently been destroyed. The Lake Waco Wetlands area provides valuable recreational and educational opportunities in addition to replacing wetlands that were lost when the level of the lake was raised.

U.S. Fish and Wildlife Service

The Brazoria National Wildlife Refuge is located east of Freeport.

City of Waco

The city of Waco provides multiple places of access to both sides of the Brazos in addition to views of the Bosque and Brazos in the city's huge Cameron Park. The Cameron Park Zoo provides the Brazos River Country exhibit, a high-quality exhibit that describes the ecology and heritage of the entire Brazos basin from the Gulf of Mexico to the High Plains. Search for "Cameron Park Zoo" and select http://www.cameronparkzoo.com/brc.html for details. Waco also has two paddling trails: the Bosque Bluffs Paddling Trail, which is 1.8 miles one way or a 3.6-mile loop, and the Brazos Bridges Paddling Trail, which is 4 miles one way or an 8-mile loop.

The City of Waco and Baylor University operate the Lake Waco Wetlands. This constructed wetland of almost two hundred acres was built to mitigate the loss of habitat when the level of Lake Waco was raised seven feet in 2003. A visitor center and display plus a floating boardwalk provide the opportunity to see wetland life and processes and to learn about research conducted by Baylor's Center for Reservoir and Aquatic Systems Research and Department of Biology. Search for "Lake Waco Wetlands" and select http://www.lakewacowetlands.com/ for specific information.

Resources for River Learning

Books

- *Aquatic Vegetation Identification Cards.* AgriLife Communications, Texas A&M System. http://agrilifebookstore.org.
- *Freshwater Fishes.* 1991. Lawrence M. Page and Brooks M. Burr. Peterson Field Guides. Boston: Houghton Mifflin.
- *Freshwater Fishes of Texas.* 2007. Chad Thomas, Timothy H. Bonner, and Bobby G. Whiteside. College Station: Texas A&M University Press.
- *A Guide to Common Freshwater Invertebrates of North America.* 2002. J. Reese Voshell Jr. Blacksburg, Va.: McDonald & Woodward Publishing.
- *Rivers and Streams.* 1999. Patricia A. Fink Martin. Exploring Ecosystems Series. New York: Franklin Watts.
- *River Wild: An Activity Guide to North American Rivers.* 2006. Nancy F. Castaldo. Chicago: Chicago Review Press.

Organizations

- Brazos River Authority Major Rivers Educational Curriculum. http://www.brazos.org/Major_Rivers.asp
- Houston Wilderness. http://www.houstonwilderness.org/defaultasp?Mode=DirectoryDisplay&id=1
- Texas Parks and Wildlife Department Nature Trackers programs, including Amphibian Watch, Box Turtle Survey, and Mussel Watch. http://www.tpwd.state.tx.us/learning/texas_nature_trackers/
- Texas Stream Team, which is part of Texas State University and trains volunteers to monitor water quality on rivers, streams, wetlands, bays, bayous, and estuaries in Texas. http://txstreamteam.rivers.txstate.edu/
- Texas Watershed Steward Program, administered by the Texas State Soil and Water Conservation Board. http://www.tsswcb.state.tx.us/managementprogram/txwsp
- Texas Nature Conservancy Columbia Bottomlands/Brazos River Project. http://www.nature.org/wherewework/northamerica/states/texas/preserves/art26883.html

Brazos River Trail

The Houston Wilderness organization promotes the preservation and appropriate uses of natural and historic areas in the Houston metropolitan area. Houston Wilderness offers the Brazos River Trail map on its Web site (http://www.houston-wilderness.org/default.asp?Mode= DirectoryDisplay&id=1). The map indicates the locations and provides contact information for the following sites on the Brazos in Fort Bend and Brazoria counties:

- Brazos Park, Community Park, and River Bend Park, Rosenberg

- George Park and George Ranch Historical Park, Richmond area
- Brazos River Park, Sugar Land
- Brazos River Park System, Missouri City
- Brazos Bend State Park, Needville
- Brazos River County Park, Angleton
- Hudson Woods Unit, U.S. Fish and Wildlife Service, near Angleton
- Varner-Hogg Plantation State Historic Site, West Columbia
- Wilderness Park, Dow Centennial Bottomlands, Gulf Coast

Bird Observatory, and Sea Center Texas, Lake Jackson
- Justin Hurst Wildlife Management Area, near Freeport
- Gulf Prairie Cemetery, Freeport
- Bell's Landing, East Columbia
- Historic Brazoria Bridge, FM 521 at Brazos River
- East Columbia Historic District, near West Columbia
- Levi Jordan Plantation, Sweeny

The Brazos River can be our teacher, if we learn to look, listen, and think.

CHAPTER 3
The Lost River

Rivers in dry regions of the world are the poor cousins in the knowledge base of river and wetland ecology. Their ecology is probably the least known of our freshwater resources.

R. T. Kingsford and J. R. Thompson, in *Ecology of Desert Rivers* (2006)

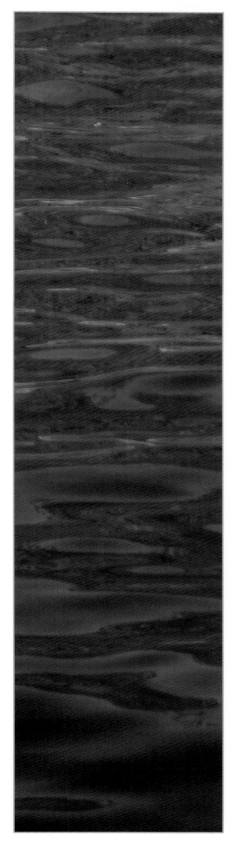

Water of the Lost River

If enough rain fell in eastern New Mexico the Brazos River would begin its flow there, on *el llano estacado,* the Staked Plains—a vast tableland at the southern tip of the Great Plains. Topographically, Running Water Draw is the headwater catchment of the Brazos. It begins north of Clovis, New Mexico, and merges with Callahan Draw to form the White River, which flows into the Salt Fork of the Brazos. Yellow House and Blackwater draws merge in Lubbock to form the North Fork of the Double Mountain Fork that emerges from Yellow House Canyon east of Lubbock. Apparently there was enough precipitation in the distant past to carve "draws" in the landscape where

water flowed, so that today's Running Water Draw, Blackwater Draw, and Yellow House Draw remain subtle features of the Llano Estacado and even carry water on rare occasions. Geologists observe that the entire Brazos River is "underfit"—today's river is not big enough to have cut its valley. They estimate that during the Late Pleistocene, which ended ten thousand years ago, rainfall in the upper Brazos was perhaps twice the current amount (Sylvia and Galloway 2006). The water that partly shaped this landscape no longer flows.

Annual precipitation on the Llano Estacado now averages less than twenty inches, and the average annual lake surface evaporation rate is more than sixty inches (Bomar 2010). Thus, most precipitation evaporates

The Lost River

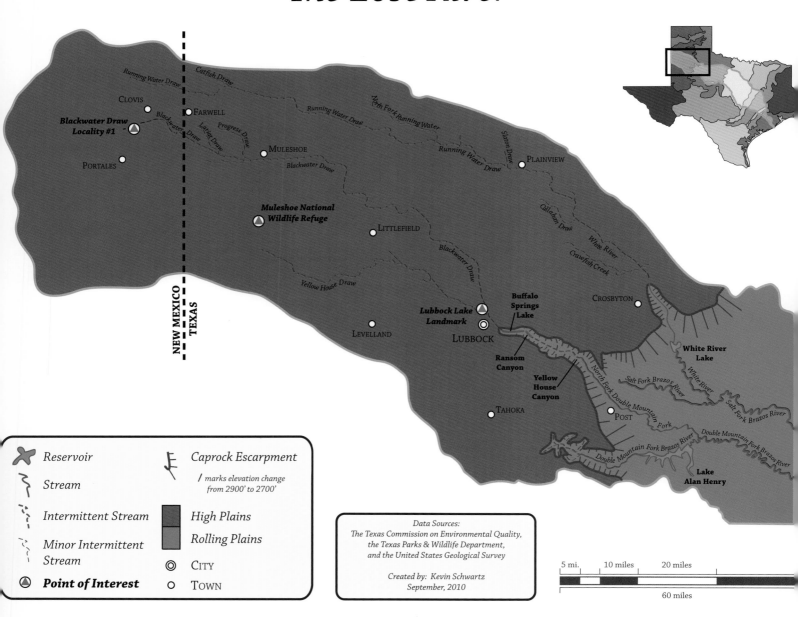

Legend

Reservoir

Stream

Intermittent Stream

Minor Intermittent Stream

▲ **Point of Interest**

⸙ Caprock Escarpment

/ marks elevation change from 2900' to 2700'

High Plains

Rolling Plains

◎ CITY

○ TOWN

Data Sources:
The Texas Commission on Environmental Quality,
the Texas Parks & Wildlife Department,
and the United States Geological Survey

Created by: Kevin Schwartz
September, 2010

5 mi.   10 miles   20 miles

60 miles

Map labels:

NEW MEXICO | TEXAS

Running Water Draw · Catfish Draw · CLOVIS · FARWELL · Blackwater Draw · **Blackwater Draw Locality #1** · PORTALES · Larita Draw · Progress Draw · MULESHOE · Blackwater Draw · Running Water Draw · North Fork Running Water · Running Water Draw · PLAINVIEW · Slaton Draw · **Muleshoe National Wildlife Refuge** · LITTLEFIELD · Callahan Draw · White River · Crawfish Creek · Blackwater Draw · Yellow House Draw · CROSBYTON · LEVELLAND · **Lubbock Lake Landmark** · LUBBOCK · **Buffalo Springs Lake** · **Ransom Canyon** · **Yellow House Canyon** · **White River Lake** · TAHOKA · North Fork Double Mountain Fork · Salt Fork Brazos River · White River · POST · Double Mountain Fork Brazos River · Salt Fork Brazos River · Double Mountain Fork Brazos River · **Lake Alan Henry**

Running Water, Blackwater, and Yellow House draws are the uppermost topographic drainages of the Brazos. They were created by flowing water in the distant past but no longer contribute to the river's flow.

Although the average annual rainfall on the High Plains is less than twenty inches, frontal storms produce heavy rains that can erode the red soil, even though the land is relatively level. Farmers plow along the contours to minimize erosion.

or is absorbed by the sandy soil. Rainfall averages are deceptive because the dominant weather system over West Texas can produce torrential storms that may pour several inches of water per hour on the land. On June 6, 1941, Running Water Draw flowed at 12,000 cubic feet per second through Plainview.

Rainfall in West Texas is primarily determined by the Marfa Front—the "Marfa Monster." Especially during the spring and early summer this front moves in a daily cycle from the west as the air heats. This "dry line" brings hot, dry air into contact with warm, moist air from the Gulf. These two very unstable air masses combine to produce steady state and severe thunderstorms with torrential rain, hail, microbursts, wind shears, and tornadoes (Bomar 1983). Such storms cause the draws to run periodically but not enough to create a river.

What little surface water is found on the Llano Estacado is in ephemeral ponds called *playas,* which may be a few hundred feet in diameter or as much as several miles. Thousands of these ponds dot the landscape, providing valuable habitat, especially for migratory birds (Sublette and Sublette 1967). Buffaloes wallowing in the dirt may have formed the playas, but more likely they were formed by a combination of wind and water action on the land surface (Jordan 1984; Sabin and Holliday 1995). Each playa is an enclosed basin and thus contributes no flow to the draws. Some of the water that collects in the playas soaks into the ground to recharge local aquifers, but most of it evaporates (Sabin and Holliday 1995).

Land of the Lost River

TOPOGRAPHY

How the Llano Estacado, the Staked Plain, got its name is lost in history, but there are many stories (Morris 1997). Some say the Spanish placed stakes as they traveled into the unknown, so they could find their way back. Others say travelers who needed to tie their horses pounded in stakes for that purpose since there were no trees. Or perhaps it was named for the yucca stalks that grow there. An elderly resident of the region told me the stakes were the remains of Indians' funeral platforms. Some geologists maintain that the Spanish explorer Francisco Vázquez de Coronado gave the region its name and that *estacado* actually meant "palisaded" rather than "staked" because the escarpment looks like palisades

(Spearing 1991). Perhaps the most interesting explanation is that *llano estacado* is a corruption of *llano estancado,* which was interpreted as "plain of many lakes" (Bolen, Smith, and Schramm 1989). However, the reason for including *llano* in the name is obvious, for the land is truly a plain, with little topographic relief and only a very slight slope of ten feet or less per mile to the southeast. The outstanding feature of the Llano Estacado is its lack of features. It's flat, in some places seemingly table flat and hence the towns there named Lamesa (The Table) and Levelland. It's not all table flat, as many writers state, although it is flatter than a

pancake; geographers calculated as much in a "roughness" comparison between a pancake and the Great Plains in Kansas (Fonstad, Pugatch, and Vogt 2003). There are gentle slopes and swales, but they're subtle. This minimal topographic relief is one reason the region contributes little flow to the Brazos. If a river is water running down a slope, this region provides little water and almost no slope. The small amount of water that does fall on the land sits there until it is absorbed or evaporates because the slight slope does not provide enough gravitational energy to make it "run." The almost constant wind accelerates evaporation.

Water on the High Plains collects in natural depressions, forming playas that provide important temporary wildlife habitat and help maintain soil moisture, even though most of the water evaporates from the shallow lakes.

SOILS, VEGETATION, AND LAND USE

The soils of the Llano Estacado are sandy loam and clay loam, deposited by water and wind from the west (Sabin and Holliday 1995). If and when there is substantial rainfall these soils are easily eroded where there is sufficient slope. The region has probably been a grassland for about eleven million years (Holliday 1990). Current native grasses include big and little bluestem, grama, buffalograss, and panicgrass (Wester 2007, 26; U.S. Department of Agriculture 2009). The landscape was maintained as a grassland by periodic grazing and fires caused by lightning and by people.

The soils are suitable for row cropping, needing only the water deep below in the Ogallala sands. By 1940 farmers had drilled almost twenty-two hundred wells in West Texas to irrigate 250,000 acres (Fite 1977). Seventy years later, 90 percent of the region was in row crops (Wester 2007).

Ecology of the Lost River

Although there is not a "river" here because there is little water or slope, there is in fact some water and some slope, enough to establish differences

The High Plains is a rich agricultural region thanks to water from the Ogallala Aquifer, not from the Lost River.

in the kinds of life found within the slightly moister places—enough for plants, animals, and people to recognize and use. The playas provide breeding habitat for a variety of birds. The playas and other surface water on the High Plains play host to hundreds of bird species, including snowy plovers, avocets, and shorebirds (Seyffert 2001). Mallards are the most common, but blue- and green-winged teal and northern shovelers are also frequent (Ray, Sullivan, and Miller 2003).

For us, the sandhill crane is the epitome of the wildness of the Llano Estacado. Sandhill cranes are big and sociable. An adult stands about four feet high, its wingspan is more than six feet, and it may weigh eleven pounds. The sandhill cranes that you can see here spend the winter on the High Plains and in northern Mexico and the summer on a breeding range that stretches from Alaska to western Quebec. The cranes migrate in flocks and gather in huge concentrations at their wintering grounds (Tacha, Nesbitt, and Vohs 1992). In 1981 about 250,000 sandhills wintered at the Muleshoe National Wildlife Refuge (U.S. Fish and Wildlife Service 2010).

These are ancient birds, one of our oldest species. Fossils of the current species (*Grus canadensis*) are about 2.5 million years old, and the genus *Grus* may be as much as 9 million to 10 million years old (Tacha, Nesbitt, and Vohs 1992). Just think about the

genetic heritage of one of these birds, going back in this very place 10 million years. We hear them before we see them when they migrate, and we are always drawn to their constant conversation as they fly. Roosting on the playas in their concentrated wintering grounds they are both noisy and smelly. Imagine 250,000 huge birds on six hundred acres of shallow water. That is slightly more than one hundred square feet per bird, equivalent to a bird almost as tall as most humans every ten feet in each direction. They are indeed sociable.

People of the Lost River

ANCIENT CULTURES

Between Clovis and Portales, New Mexico, in an area of rich farmland irrigated by water from the Ogallala Aquifer, is one of the most important sites in the archeology of the New World. Here in 1929 the first characteristic stone projectile points that came to denote the Clovis culture were found. The site is in Blackwater Draw, which is one of the non-contributing tributaries of the Brazos. Long considered to be the earliest human culture in the New World, Clovis people left traces at sites found virtually all over the United States, including along the Brazos. At one time archeologists theorized that the Clovis people were completely nomadic specialized hunters who preyed on the megafauna of the Pleistocene, even

Sandhill cranes at Muleshoe National Wildlife Refuge connect us with the ancient past and hopefully cause us to think about a sustainable future.

years ago to today's climate beginning about forty-five hundred years ago. During that time it experienced two severe droughts, each lasting several hundred years (Holliday 1987). During the droughts there was only a little brackish water in the draws, but it supported life. Even during the drought times the area was not abandoned (Holliday 1987). Of course, there were no permanent settlements here and very few people.

Blackwater Draw Museum

http://www.enmu.edu/services/ museums/blackwater-draw/ museum.shtml

The Blackwater Draw Museum is located at 42987 Highway 70, about seven miles northeast of Portales and eight miles south of Clovis, New Mexico. The site of the original and ongoing Clovis excavation is at the Blackwater Locality No. 1 Site, a few miles from the museum. The museum and site are operated by Eastern New Mexico State University. It is possible to tour the site with permission from university staff.

Artifacts of the Clovis culture, often considered to be the first in the succession of human cultures in North America, were first identified in Blackwater Draw, where excavation continues at the Blackwater Draw Museum. Artifacts of the Folsom culture are also found in Blackwater Draw.

to the extent that they caused many of those animals to go extinct, including the woolly mammoth and mastodons. However, newer interpretations based partly on data from Clovis sites in other parts of the Brazos watershed have initiated a re-interpretation of how the Clovis people lived.

There was a long succession of people here. Using differences in stone points as criteria, archeologists traditionally divide the past dozen millennia in the far upper part of the Brazos into the Paleoindian period (11,000 to 8,500 years BP), the Archaic period (8,500 to 2,000 BP), the Ceramic period (2,000 to later than 1,000 BP), the Protohistoric period (about 500 to 300 BP), and the Historic period (beginning about 300 BP), but this chronology is now being revised (Holliday 1987; Bousman 2004). The area changed from a mild, relatively moist climate eleven thousand

BUFFALO HUNTERS AND RANCHERS

The grasslands of the Llano Estacado supported grazing animals for millions of years, including many of the 100 million bison of the larger Great Plains region. American hunters had almost eliminated the buffalo by the 1880s. The Comanche, who depended on the buffalo and dominated the plains, were consequently defeated (Wester 2007). New grazers moved in, the first being Mexican sheep with their herders, who were soon replaced by cattle and ranchers.

The huge buffalo population roamed the grasslands for eons, grazing heavily and then moving on. They did not return to a specific site for years, thus allowing the grass to regenerate. However, cattle were confined by their owners to ranges where they might graze almost continuously. They quickly overgrazed the range, making it necessary to allow fifty acres per head where previously five acres would support one animal. Droughts and blizzards in 1885 and 1887 killed millions of cattle, and the ranching industry never again produced the numbers it had prior to those events (Wester 2007).

In spite of the disdain it receives from some people, properly managed grazing is probably the most sustainable way we have to produce food and other products. Proper grazing consumes only about 25 percent of the forage, leaving the rest for insects and other native grazers, which continue to add organic matter to the soil. The disaster of grazing in much of the Southwest actually resulted from *over*grazing.

FARMERS

After the range deteriorated from overgrazing the region fell under the delusion that the plow would bring rain. Land speculators drilled a well and filled a playa near the train station at Plainview to convince prospective buyers of the abundance of water (Bolen, Smith, and Schramm 1989). Today the region is one of the most intensively cultivated areas in the United States. Cotton is the dominant crop: 15 percent of the cotton grown in the United States and 3.5 percent of the cotton grown worldwide is produced on the South Plains (Wester 2007). Although the plow did not bring rain, it motivated development of sophisticated irrigation systems that extract water from the Ogallala Aquifer with powerful pumps that can draw up to one thousand gallons per minute.

As with most aquifers, the Ogallala naturally recharges from precipitation, but the precipitation is limited and withdrawals are usually much greater than recharge. Thus, the aquifer continues to decline. However, farmers and agencies that work with them are exploring and incorporating new methods of irrigation and tillage to conserve the precious water that is the basis of their very lives on the High Plains.

CITY CEOPLE

The city of Lubbock was founded in 1890 as an agricultural trade center and the seat of Lubbock County. It continued to develop its agricultural and commercial trade after the Santa Fe Railway provided service in 1909. In 1923 the Texas Legislature authorized the establishment of Texas Technological College in Lubbock. Later renamed Texas Tech University, it became one of the state's most important universities (Graves 2009).

The Lubbock County metropolitan area had an estimated population of 270,550 in 2009, most of that population residing within the city of Lubbock (U.S. Census Bureau 2009). Currently the city obtains 80 percent of its water from the Canadian River Municipal Water Authority, which supplies water from Lake Meredith, a reservoir on the Canadian River, and from wells drilled into the Ogallala Aquifer. Lubbock draws the other 20 percent of its water from its own wells in the aquifer (City of Lubbock 2008). However, both of these sources are limited and/ or threatened. In January 2010 Lake Meredith was at a record low and the Ogallala was declining one to three feet per year (Blackburn 2010; Canadian River Municipal Water Authority 2008). Since these water sources are not directly related to the Brazos River, they are not immediately part of our story. However, Lubbock owns the relatively new Alan Henry

Reservoir, built in 1993 below the Caprock southeast of the city on the Double Mountain Fork of the Brazos River. We will consider the effects of this water storage reservoir in the next chapter.

Where to Experience the Land and the River

MULESHOE NATIONAL WILDLIFE REFUGE

Because the eastern High Plains are now almost completely planted in irrigated row crops there are few natural areas and even fewer places where the public has access. However, the Muleshoe National Wildlife Refuge provides relatively natural habitat and public access. Twenty miles south of Muleshoe, Texas, and dating to 1935, this 5,809-acre refuge was the first national wildlife refuge established in Texas. Its primary purpose is to provide natural wildlife habitat, especially for migratory birds. The U.S. Fish and Wildlife Service owns and manages the refuge, using both cattle and controlled fire to replicate the natural effects of bison and natural fires. There are three lakes on the refuge that provide about six hundred acres of water when full. One lake is spring fed, but the other two are true playas that only receive water from runoff, so they are full only after rains (U.S. Fish and Wildlife Service 2010).

Prairie dogs, cottontails, and jack-rabbits are common and easily seen at the refuge. Coyotes, bobcats, badgers, and even porcupines are there as well. The refuge lists 320 bird species, ranging from mourning doves to golden eagles, but the signature bird of the refuge is the sandhill crane. The sandhill cranes begin to arrive at Muleshoe in late September and early October. Their population peaks between mid-December and mid-February (U.S. Fish and Wildlife Service 2010).

The Muleshoe National Wildlife Refuge offers primitive camping, a picnic area, and nature trails. It is located south of Muleshoe, Texas, on State Highway 214.

LUBBOCK LAKE LANDMARK

Blackwater Draw and Yellow House Draw converge at what is now called the Lubbock Lake Landmark, on the northwest side of Lubbock. This is the best place for us acquire an understanding of the human importance of draws on the High Plains. The draws, with their scant but important water, served as travel routes through the High Plains for ancient people. Archaeologists have studied the rich resources at Lubbock Lake since the 1930s, so there is a tremendous amount of information about the geology, archaeology, and history of the site. Lubbock Lake Landmark is open to the public.

In a strange synchronicity, my father's cousins, Clark and Turner Kimmel, were teenagers when they found an unusual stone point at Lubbock Lake in 1936. I would like to think that it was due to our family's strong commitment to intellectual pursuits that the boys gave the point to an archeologist at the college, but more likely they did so because someone wiser understood its value. However it happened, their find initiated the process that recognized the archeological importance of the site.

In the late nineteenth century the site was a natural, spring-fed lake of about ten acres. George Singer built the first store in Lubbock County near the lake in the early 1880s and cattlemen watered their stock there. Residents realized that the dry land was underlain by the huge Ogallala Aquifer, and they quickly developed irrigated agriculture. Pumping this precious groundwater dried up the springs by the early 1930s. The City of Lubbock, with funding from the Works Progress Administration, dug out the springs and a reservoir basin to make a lake that would provide water for fire protection (Johnson and Holliday 1987). Clark and Turner Kimmel found the stone point while wandering around the excavations.

Vance T. Holliday, who, along with Eileen Johnson, conducted and directed much of the archeological work at Lubbock Lake, has stated that "the outstanding stratigraphic record and age control at Lubbock Lake provide an excellent physical and temporal framework for establishing a comprehensive, local cultural

chronology encompassing the last 11,000 years of human occupation" (Holliday 1987, 22). What can this history of the past twelve thousand years on the High Plains tell us about this place and, most importantly, about ourselves and the future?

I sit on a slope overlooking the Lubbock Lake site and my mind wanders back to my father's cousins finding what they likely called an arrowhead. Rather than an arrowhead, it probably was from a short spear, propelled by a throwing stick or *atlatl*, which gave it great power at a short distance. The Texas Parks and Wildlife Department commissioned full-size bronze statues of the mammoth and other animals whose remains were found at the site. I look at those statues and think about the danger in trying to kill one of those animals with a spear.

The lesson for us from this period in the upper Brazos is that life is resilient as long as there is a little water, even water that is barely usable. The Lubbock Lake site never completely dried up, and people never abandoned the place. Did they mourn the good old days? Perhaps their legends spoke of better times, but they each dealt with their own reality, as we

Playa lakes of the High Plains provide habitats for a wide variety of life, especially migrating waterfowl. At the Muleshoe National Wildlife Refuge those habitats are maintained and are accessible to the public.

Life-size sculptures of prehistoric animals spur the imagination about what life was like for the hunting and gathering people who lived at what is now the Lubbock Lake Landmark, where Blackwater Draw and Yellow House Draw converge.

all do. Did young people court each other and make babies? Obviously. Did they mourn their dead? Surely, because death was frequent. Was there joy and hope? We have no way to know, but a fresh bison kill or a fish in a trap must have brought the anticipation of pleasure and hope for the next kill or catch.

The Lubbock Lake Landmark site is operated by Texas Tech University. The site offers guided and self-guided tours, a learning center, and

exhibitions. It is possible to watch archeologists at work during the summer when excavations are under way. The landmark is listed on the National Register of Historic Places and is a National Historic and State Archeological Landmark.

The Lubbock Lake Landmark is located on the north side of Lubbock on Landmark Drive, near the intersection of North Loop 289 and U.S. 84.

CHAPTER 4

Many Arms of God

There are no great mountains, no great lakes and no great rivers in this land of mine. There are a few big creeks, some of which we designate as rivers, and they become mighty rivers at flood stage, able to float the largest steamer on the Mississippi. It is a land of broad prairies, quiet valleys and vast distances: a land of bright skies, glorious sunsets and most brilliant starlight; a land where the hills and plains are gay with lovely wildflowers in the springtime, and where we have the everlasting hills unto which we may lift up our eyes. It is a land where the cries of the coyote and the hoot-owl and an occasional scream of a panther break the nocturnal stillness.

From *Interwoven: A Pioneer Chronicle* (1936), by Sallie Reynolds Matthews, born in the upper Brazos country in 1861

Water of the Many Arms of God

Things haven't changed much since Sallie Reynolds Matthews published those words in 1936. Her descendants still own some of the land that is cut by the four streams that ultimately form the Brazos River. The major change is that oil has brought wealth to the region. In spite of the wealth, or perhaps because of it, many people in the region are dedicated to honoring their more humble heritage.

The Spanish name for our river is Los Brazos de Dios, "The Arms of God." Historians relate a number of legends accounting for the name, ranging from Coronado being saved from thirst or from pursuing Indians in the headwaters to thirsty Spanish sailors finding fresh water at the

Los Brazos de Dios (clockwise from upper left): the North Fork of the Double Mountain Fork, the Salt Fork, the Clear Fork, and the Double Mountain Fork.

The Many Arms of God

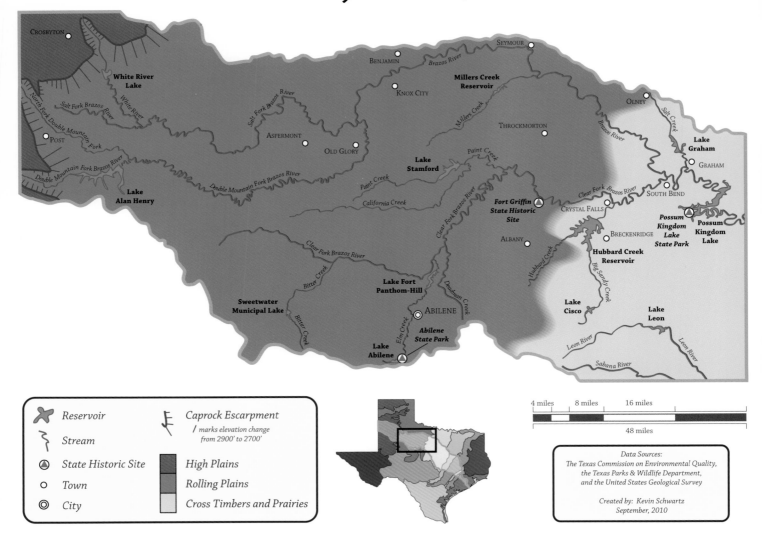

Reservoir	Caprock Escarpment / marks elevation change from 2900' to 2700'
Stream	
State Historic Site	High Plains
Town	Rolling Plains
City	Cross Timbers and Prairies

4 miles   8 miles   16 miles

48 miles

Data Sources:
The Texas Commission on Environmental Quality,
the Texas Parks & Wildlife Department,
and the United States Geological Survey

Created by: Kevin Schwartz
September, 2010

Stream (time period)	Average annual flow (cfs)	Range of annual flows (year)	Peak flow (date)
Double Mountain Fork near Justiceburg, upstream from present Lake Alan Henry (1962–2007)	28.7	1.65 (1983) to 75.6 cfs (2005)	49,000 cfs (May 6, 1969)
Salt Fork near Aspermont (1964–2007)	63.3	11.7 (1998) to 212.3 cfs (1987)	52,200 cfs (September 25, 1955)
Clear Fork at Fort Griffin (1939–2007)	198.8	8.78 (1952) to 1,177 cfs (1957)	149,000 cfs (August 4, 1978)
Brazos River near South Bend, downstream from confluence with the Clear Fork (1962–2007)	686.0	174.4 (2000) to 2,966 cfs (1992)	87,400 cfs (May 4, 1941)

Source: U.S. Geological Survey National Water Information System Web Interface USGS Surface-Water Annual Statistics for the Nation
http://waterdata.usgs.gov/nwis/annual/?referred_module=sw (accessed March 20, 2006).

mouth. It is possible that early map makers confused this river with the Colorado (Hendrickson 2008). The name Brazos de Dios is particularly appropriate for the upper part of the drainage, where four different streams containing the name "Brazos" come together to form the mainstem of the Brazos River: the Salt Fork of the Brazos, the North Fork of the Double Mountain Fork of the Brazos, the Double Mountain Fork of the Brazos, and the Clear Fork of the Brazos.

WATER FLOW

The official Texas highway map shows the four upper tributaries that form the Brazos in a nice watery blue. But the water is not blue; it is tan to bright red. The average annual rainfall in the region is twenty-two to thirty inches, enough to produce substantial flows, especially given the high-intensity storms that are common. Usually the flow is low, although flooding is frequent and often high. The sidebar shows the wide variations between average annual flows, the range of annual flows (low and high), and the single peak flow for each of the four tributaries. Without getting overly caught up in numbers, let's consider the Clear Fork. Its long-term average flow is almost 23 times greater than its lowest annual flow. Its highest annual flow is about 6 times its long-term average. But most astounding is its peak flow, which is 750 times greater than the long-term average flow. It doesn't

The red soils drained by the upper Brazos are derived from Permian deposits and are easily eroded, giving the river an almost unreal color, as seen here east of Newcastle.

matter what the averages are; it's the big floods that wash away houses, bridges, livestock—and people.

If you are a numbers person you may note what appears to be a mistake in the table. The peak flow recorded on the Clear Fork was 149,000 cfs, but the peak flow of the Brazos downstream, which receives the flow of the Clear Fork, was only 87,400 cfs, and did not occur at the time of the big flood on the Clear Fork. It would seem obvious that a big flood upstream would lead to a big, maybe even bigger, flood downstream. Actually, two days after the big flood on the Clear Fork, the Brazos downstream at South Bend flowed at 78,100 cfs—high, but substantially less than the flow of its upstream tributary. The reason for this anomaly is that the high-intensity storms of the region tend to be localized, so the downstream area did not receive as much rain. In addition, the channels and floodplains of dryland rivers have a high capacity to absorb or store water, thus reducing flooding downstream (Young and Kingsford 2006).

We noted in the previous chapter that the city of Lubbock has expanded its reach for water into the Rolling Plains. With its pretty name and an average annual flow of only 28.7 cfs, the little Double Mountain Fork doesn't seem like much of a river. It is the smallest by far of the Arms of God. But that seemingly small flow adds up to an average of 18,549,295 gallons per day, now mostly retained behind Alan Henry Dam, minus evaporation and absorption losses. The Alan Henry Reservoir was completed in 1993 as a water supply for Lubbock. The lake's surface area is 2,880 acres when full, with depths up to 100 feet (Texas Parks and Wildlife Department 2008a). Lake Alan Henry is long and narrow, with a rocky sandstone shore. The water is clear, and the lake is an important attraction in this generally dry region.

The business of marketing water as a commodity may yet affect the hydrology of the region and the Brazos in particular. Operating a company called Mesa Water, oil entrepreneur T. Boone Pickens organized a number of landowners to drill wells on their properties on the High Plains and sell that water to thirsty Texas cities. The following excerpt from the Mesa Water Web site explains the rationale:

> Mesa Water represents a group of Texas Panhandle landowners, led by Boone Pickens, who put a lot of stock in two basic things . . . land and family.
>
> For generations, these families have lived and worked in the rolling hills they love. God blessed their land with an underground aquifer filled with naturally pure groundwater. And thanks to the Ogallala Aquifer, these landowners have more water than they can ever use.
>
> With the population of Texas booming and a perpetual drought predicted to hit our area as soon as 2021, water planners [and] state and local leaders are looking everywhere for a solution. They know the key to secure, drought-proof, long-term water planning is diversity . . . building reservoirs, encouraging conservation, capturing and purifying runoff and buying water from another region and piping it to where it is needed.
>
> Today, Mesa Water is ready to sell water to communities that don't have enough for the future. (http://www.mesawater.com)

The company's business plan called for selling as much as 320,000 acre-feet of water per year, an amount estimated to meet the needs of 1.5 million people. (Acre-foot is the most common measurement of large volumes of water. It is the amount of water that will cover an acre of land to a depth of one foot, an amount that is equal to 325,851 gallons.) Depending on the customers, Mesa's plan involved transporting the water through the channels of the Trinity and Brazos rivers. From the scattered wells, the water would flow by pipe to the Brazos near Graham, Texas, and then the river channels would be used to convey the water to Mesa's customers. Mesa Water found no buyers, so the project is on hold, but Texas is a thirsty state and the proposal could arise again.

WATER QUALITY

The environmental historian Dan Flores offers a provocative perspective about water quality in the upper Brazos: "Yellow House Canyon's string of treated sewer-water lakes in the draw above the canyon . . . is no doubt an improvement over the industrial dump grounds that were once there, although it might give pause to residents downstate to realize that the Brazos River now has its 'headwaters' in flushing toilets in Lubbock" (Flores 1990, 174).

Actually, today many rivers in Texas depend on the "return flow" of treated wastewater effluent. A properly operated wastewater treatment plant reduces the amount of harmful bacteria to levels deemed to be acceptable. However, many wastewater treatment plants are not required to reduce the amounts of nutrients such as nitrogen and phosphorus, which stimulate growth of aquatic vegetation, including algae and bacteria. Accelerated growth of algae and bacteria can cause water pollution. Photosynthesis by algae produces oxygen during the day, but respiration at night and decomposition consume dissolved oxygen in the water, thus reducing the level of this very important component of a healthy aquatic system.

SALT

The water faucets in the public facilities at Possum Kingdom State Park are labeled "Not Potable" because the water in Possum Kingdom Lake is too salty for human consumption. In fact, the Brazos all the way to Lake Whitney exceeds state standards for chlorides in drinking water. Most of the salt comes from the aptly named Salt Fork of the Brazos, especially from its tributary, Salt Croton Creek. At the monitoring station at U.S. Highway 83 the Salt Fork has contained as much as 68.6 parts per thousand (ppt) of salt, a level that is almost twice the average salt content of seawater. Chloride concentration averages 28,987 milligrams per liter (mg/L), compared to the state drinking water standard of 300 mg/L, and total dissolved solids average 44,779 mg/L, compared to the standard of 1,000 mg/L. Stated another way, this part of the Brazos contributes about 14 to 18 percent of the total flow of the river but produces about 45 to 55 percent of the total dissolved solids, 75 to 85 percent of the chloride, and 65 to 75 percent of the sulfate measured in the Brazos at Richmond, near the river's mouth (Brazos River Authority 2008).

Reading between the lines in the following long quotation from the Brazos River Authority (2008), it is possible to detect the frustration of an agency that is expected to "manage" the river for human benefit:

Since the mid-1930s the Brazos River Authority has worked with numerous agencies to research the sources of salt pollution, methods of control and abatement, as well as costs-benefit analyses of potential salt removal solutions throughout the Brazos River Basin. Chloride compounds occur in almost all natural waters and sources can be of natural origin, dissolved from minerals, contamination from seawater or from wastes. The high concentrations of natural salt enter the waters of the Brazos River in the Upper Basin Region northwest of the City of Abilene, principally in the watersheds of the Salt and Double Mountain Forks of the Brazos River. Croton and Salt Croton Creek contribute a substantial part of the salt load in the Brazos River. The natural salt pollution producing area is a semi-arid region of salt and gypsum encrusted hills and canyon-like stream valleys. The area is studded with salt springs and seeps. The highly erodible floodplain material in this region is continually washes [sic] away as the streams cut their way down to rock or other impervious base. The bedrock provides a cap over a brine aquifer that underlies the entire region of Texas and parts of Arkansas, Oklahoma, and Kansas. In an area where the erosion process has continued for

River segment	Water quality concerns	Causes
North Fork of Double Mountain Fork	Elevated bacteria levels, nutrient enrichment, and algal growth	Stormwater and municipal discharges from Lubbock area and wildlife wastes
Clear Fork	All segments meet water quality standards, but one tributary shows nutrient enrichment from wastewater flow.	Most Texas wastewater treatment plants are not required to reduce nutrient content.

Source: Brazos River Authority. 2007. Clean Rivers Program Basin Summary Report. http://www.brazos.org/BasinSummary_2007.asp (accessed November 14, 2007).

You can fly Google Earth® to 33° 21' 24" N and 100° 30' 02" W at an eye altitude of fourteen miles and see the salt flats. Zoom down to see how salt is deposited down the streambed.

As a people we often do not accept nature as it is, so we're still trying to fix the salt "problem" in the Brazos. Maybe we shouldn't, as some biologists would maintain, because the river's ecosystem has evolved in an environment with a high salt content. But probably we will try and keep trying until we succeed or run out of money. Many people think the water is too valuable to leave in its natural, salty condition. The Possum Kingdom Water Supply Corporation in 2006 received a loan of $4.7 million from the Drinking Water State Revolving Fund of the U.S. Environmental Protection Agency, plus $6.5 million in Rural Development Funds from the U.S. Department of Agriculture, to purchase and consolidate small water supply systems that did not meet state standards for chlorides, sulfates, and total dissolved solids (U.S. Environmental Protection Agency 2007). The corporation also received authorization to issue nearly $1.63 million in revenue bonds to increase its reverse osmosis desalinization capacity from 1 million gallons per day (mgd) to 1.5 mgd. The reverse osmosis process uses electricity to power pumps that force water through a membrane, thus separating water from the salts. However, only 25 to 50 percent of the water is usable and the remaining

Natural salt deposits in the Salt Fork raise the salinity of the Brazos so much that the levels exceed state drinking water standards as far downstream as Lake Whitney.

centuries, the streambed has spread out to form large flats. Wherever there is a fracture joint in the stream bedrock material, the highly mineralized water seeps to the surface under artesian pressure; massive salt flats, often 400 to 500 acres in size, are formed by this process. Salt and other minerals also leach out of the adjacent floodplain material that surrounds the salt flats and streams. The Brazos River receives a tremendous salt load when local rainfall is sufficient to dissolve the deposited salt and washes it out of the salt flats.

"waste" water has an even higher salt concentration.

Decades of high-volume pumping from the Ogallala Aquifer have diminished spring flow below the Caprock Escarpment and thus reduced the baseflow of the streams that originate below the Caprock (High Plains Water Conservation District No. 1 2008). It is likely that the diminished baseflow of fresh water has increased the salinity of the salt streams.

Earthen "tanks" are the primary water supply for cattle where groundwater is too salty.

IMPOUNDMENTS

If you fly to Albany, Texas, on Google Earth® at an eye elevation of about eighteen thousand feet, you will see two outstanding features on the land. The most obvious is the large number of white spots connected to a network of white lines. Those are oil well pads and gravel roads, graphically demonstrating the basis of the wealth of the region.

The other feature you will see are round or irregularly shaped spots or blotches of green, black, or brown. Those are water impoundments called "tanks." There are literally

thousands of them in the Brazos watershed. Because much of this region is used for ranching and there is little usable water available for the cattle, ranchers usually provide a "watering" so that cattle do not have to walk more than a mile to drink, in order to prevent weight loss. In many ranching areas these "waterings" are supplied by wells pumped by the classic western windmills. However, the groundwater in this section of the Rolling Plains is too salty, so ranchers build ponds by scooping out a depression in a small drainage way and using the material that is removed to

make a dam on the downstream side of the depression.

As you fly around the region you will also see many larger "lakes" built on small tributary creeks to supply water to the communities, again because the groundwater is too salty for human consumption. The region currently has a population of about 270,000 people, half of whom live in the Abilene area. Abilene relied for years on Lake Abilene and Fort Phantom Hill reservoirs on the Elm Creek tributary of the Clear Fork. The city now pumps water from Hubbard Creek Reservoir, more than fifty miles away, near Breckenridge, and from O. C. Ivey Reservoir, more than sixty miles away, in the Colorado River basin. Thus, although the region does not seem to "use" the upper Brazos River tributaries directly, it depends greatly on the surface water of the Brazos catchment basin.

Under the Conservation Amendment of 1917 the State of Texas authorized local water districts to have "unlimited borrowing and taxing power to conserve and develop water resources" (May 2008). The purpose of the districts was to provide water for domestic uses. The table listing dams and reservoirs in the Rolling Plains section of the Brazos River basin shows that a number of cities, including Abilene, Cisco, Mineral Wells, and Sweetwater, quickly took advantage of this legislatively granted financial power to develop their water supplies. All of the reservoirs in the region were built for water supply rather than flood protection or hydroelectric power. The average age of the reservoirs listed in the table is about sixty-three years, and five of them are more than eighty years old. Reservoirs fill with silt over time and often are not expected to be useful for more than one hundred years, but sedimentation is determined by land use practices, watershed characteristics, and climate (Wetzel 2001). Some of the cities of the region are seeking additional water supplies.

Land of the Many Arms of God

TOPOGRAPHY

The Rolling Plains extend from the Caprock Escarpment east of Lubbock to just upstream of Possum Kingdom Lake. The elevation varies from about three thousand feet at Buffalo Springs Lake in Yellow House Canyon to about one thousand feet at Possum Kingdom Lake. "Rolling Plains" is an appropriate name because the landscape is rolling, with some moderately rough topography cut by narrow stream valleys running generally southeast. The "rolling" nature of the land is both the cause and result of water's erosive power, as the headwater tributaries erode their way westward.

SOILS, VEGETATION, AND LAND USE

The soils of the Rolling Plains are easily eroded sand and clay loam that

Rolling Plains section dams and reservoirs	Date built	Original capacity (acre-feet)	Surface area (acres)
Abilene Lake	1919	6,099	595
Alan Henry Reservoir	1993	94,808	2,880
Buffalo Springs Lake	1960		241
Cisco Lake	1923	26,000	1,050
Daniel Lake	1948	9,435	950
Davis Lake	N/A	5,454	
Fort Phantom Hill Lake	1937	70,030	4,213
Graham Lake	1958	45,260	2,444
Hubbard Creek Reservoir	1962	318,067	14,922
Kirby Lake	1928	7,620	740
Leon Lake	1954	26,421	726
Millers Creek Reservoir	1974	27,888	2,212
Mineral Wells Lake	1920	7,065	646
Palo Pinto Lake	1965	27,150	2,399
Ransom Canyon Lake	1960s	private	
Stamford Lake	1953	51,570	5,124
Sweetwater Lake	1929	10,066	630
White River Lake	1963	29,880	1,418

support a variety of grasses and upland woody vegetation. About one-third of the land area is used for row crops and two-thirds are mesquite grasslands. There is little irrigation in this region in comparison to the High Plains to the west, where water from the Ogallala Aquifer supports agriculture. The Seymour Aquifer underlies the Rolling Plains in patches and provides some irrigation water, but it has a high salt content due to the same natural salt deposits that raise the salinity of the Brazos (Texas Water Development Board 2008). Row cropping here exposes the sandy soils to erosion, thereby increasing the sediment load in the Brazos and producing its bright red color.

Ecology of the Many Arms of God

The four rivers that make up the Brazos de Dios are small and have widely variable flows. Their channels are generally deeply cut into the erodible landscape, which supports thick riparian vegetation, including willow, pecan, and elm trees. The streambeds and banks consist of sand and finer particles. The rivers carry a substantial load of fine sediment. As described earlier, the Salt Fork has very high dissolved mineral content.

These four upper tributaries that form the Brazos correspond well with the river continuum concept outlined in chapter 2. Most of the organic matter that supports life in the river comes from the surrounding land, especially from riparian vegetation. Photosynthesis is limited by shading from the streamside vegetation and by the particles in the water (turbidity) that reduce the penetration of sunlight needed for photosynthesis.

Standing on a bridge near Fort Griffin we see twirling water bugs, carp, four water snakes, and a school of at least twenty gar nose-first in the slight current across a sandbar. We don't see the catfish that anglers seek, nor the young mountain lion that has been spotted here. We have seen the feral hogs that are a scourge on the land. We see and hear an amazing variety of birds, from painted buntings to owls.

Biologically, the Salt Fork of the Brazos is the most interesting of the four tributaries due to its naturally high salinity. Over the eons species have evolved to live successfully in the high salinity. The most notable of those is the minnow-size Red River pupfish (*Cyprinodon rubrofluviatilis*), which has been found in water with salinity as high as 150 parts per thousand, compared to the average 35 parts per thousand of seawater (Echelle, Hubbs, and Echelle 1972). Red River pupfish can also tolerate low oxygen content and high water temperature, so they are well suited to live in the shallow, salty water of the Salt Fork.

Although not as biologically unique as the pupfish, two other species seem to epitomize highly adapted life forms in and along this part of the river—longnose gar and salt cedar: the gar because it is such an ancient and resilient fish and salt cedar because it is an aggressive newcomer.

LONGNOSE GAR (*LEPISOSTEUS OSSEUS*)

Gars are ancient fish, dating from the Permian period more than two hundred million years ago. They are predators, eating smaller fish and invertebrates. They are tough and bony; both the species name and part of the genus name of the longnose gar mean "bony" in Latin and Greek, respectively. Longnose gar can be as long as six and a half feet and weigh up to fifty pounds. The record longnose gar from the Brazos was fifty-six inches long and weighed twenty-four pounds (Texas Parks and Wildlife Department 2008b). Alligator gar, which are common farther downstream, can be much larger.

The gar's capability to survive for so long in difficult and changing habitats is because unlike most fish, which absorb oxygen solely through gills, gars are also able to gulp air. Like most fish, gars have an air bladder that helps them maintain buoyancy. But the gar's air bladder contains blood capillaries that allow it to function as a lung. Thus, the gar can survive in water with low oxygen content where other fish would not be able to live (Renfro and Hill 1970).

Ancient animals like gars put all of life in perspective. It is astounding to

know that their form and being, their genetic code, are hundreds of millions of years old, that the ancestors of the very fish we watch from the bridge have been in this river for its entire existence of perhaps ten million years.

SALT CEDAR (*TAMARIX* SPP.)

Salt cedar (tamarisk) is also an ancient species, but it is new to North America and not welcome even though we brought it here. Tamarisk is an Old World tree brought to the New World in 1854 to create windbreaks and shade, to stabilize streambeds, and to serve as an ornamental (Hart et al. 2005). It has spread widely through stream courses throughout the southwestern United States. Its biology accounts for its success—it can find water, tolerate fire, and thrive in salty soils. The plants have long taproots that draw water from deep in the soil. If the groundwater is salty the salt cedar takes in the salt, concentrates it in its foliage, and then drops the salty leaves on the ground, thus retarding other plant growth. In addition to producing seeds, salt cedar reproduces by sending shoots up from its roots. The flowers produce thousands of tiny seeds, each with a

The soils derived from the red deposits of the Permian period support a variety of agriculture and in many places are starkly beautiful. Of course, oil wells are a favored crop.

The aquatic habitats of the upper tributaries are somewhat harsh, with high levels of dissolved and suspended solids and widely fluctuating flows and temperatures, but life adapts.

tuft of hair to fly on the wind. Seeds can also be dispersed by water (U.S. Forest Service 2008).

Salt cedar crowds out native vegetation and draws huge amounts of water from streams. For example, a square yard of a typically dense stand of salt cedar may transpire about two hundred gallons per day, removing the water from the river and releasing it to the atmosphere (Hart et al. 2005).

Salt cedar is very difficult to control. Since it is able to reproduce via

Thick stands of tamarisk (salt cedar) choke much of the riparian zone of the upper tributaries of the Brazos.

its roots, cutting it is ineffective as well as very expensive. Herbicides are also expensive and may harm other vegetation. The latest control effort involves the introduction of an insect called the salt cedar leaf beetle (*Diorhabda elongata Brulle*). This beetle feeds on salt cedar leaves and can completely defoliate a tree (U.S. Army Corps of Engineers 2007). Studies of native plants potentially affected by the beetle have so far indicated that it is safe to release (Dudley and Kazmer 2005). Use of biological controls such as the salt cedar leaf beetle requires more knowledge of ecosystem characteristics and may have a slower effect than herbicides, but the benefits last much longer (Eberts n.d.).

People and the Many Arms of God

As we will see later, the kinds of people who camped at Lubbock Lake in the distant past also traveled and camped along the streams in the middle and lower parts of the Brazos. Although there is little archeological evidence of their presence, we might assume these people and others followed the river through this region as well because in it they could find water and food, both much harder to find in the uplands away from the river.

In the Canadian River valley, which is the second river basin north of the Brazos, archeologists have documented a culture called the Plains Village tradition. These people were hunters and gardeners who built stone and adobe houses between about AD 1100 and 1500 and mined chert (flint) at what is now the Alibates Flint Quarries north of Amarillo (Texas Beyond History 2010). Although archeologists have not reported Plains Village sites on the Brazos, the hunting-horticulture lifestyle of the people would be possible in the valleys of the upper Brazos tributaries.

A more nomadic group in the region between 1500 and 1700 was called the Jumanos (Hickerson 2009). The Plains Village tradition and the Jumanos were overcome by Athabaskan peoples who emigrated from the north. These people became the Navajo to the west and the Apache to the east, apparently arriving in the region shortly before the Spanish explorers (Carlisle 2009). By the mid-1700s the Apaches were dominated by another group of newcomers. That group came to be known as the Comanche, a name derived from the Ute word *komántcia*, meaning "anyone who wants to fight me all the time" (Lipscomb 2009). The Kiowa may have evolved from some of the Jumanos and learned the horse and hunting skills of the Comanche (Hickerson 2009). The Kiowa and Comanche cooperated to dominate the region (Texas Beyond History 2009b).

During this shifting power struggle, Coronado may have crossed the upper Brazos tributaries seeking Quivira in 1541. The remainder of Hernando De Soto's expedition possibly camped in about 1542 on the Double Mountain Fork of the Brazos, which they called Daycao (Stephens and Holmes 1988), but not all historians agree (de la Teja, Marks, and Tyler 2004). Even if they did visit, the Spanish explorers found nothing of interest and did not return to this part of the Brazos.

The Comanche and Kiowa Indians fiercely protected their buffalo hunting grounds. The Rolling Plains of Texas, through which our Many Arms of God pass, was the eastern margin of the range occupied by the Great Plains buffalo herds. Thus, the region was the meeting ground between Indians, settlers moving in from the east, and gold seekers on their way to California (Texas Beyond History 2009b).

The life strategy of both the Indians and the whites in the region was determined by the characteristics of the semiarid prairie land. Indians relied on the grazing buffalo and increasingly on raiding. Whites based their livelihood on cattle and thus depended on the grass just as the Indians did. Indians based their resistance on hit-and-run tactics made possible by the horse and an intimate knowledge of the open country. Whites responded with similar tactics, although they had to learn them by experimentation because most of their military-type tactics were based on the ancient European method of matching wave after wave of men on foot in battles of attrition.

From the 1840s to the early 1880s there were continual running battles between these groups that had such different expectations of life and land. To protect the settlers the U.S. Army built a line of forts in 1849 from Fort Worth to the Alsatian-German settlements west of San Antonio (Texas Beyond History 2009b).

Within two years the military realized that the 1849 line of forts was not effective, so the army established another line of forts farther west. This new group included Fort Belknap on the Brazos, near the present town of Newcastle, about eighty miles west of Fort Worth. What is now known as Fort Phantom Hill was built on the Clear Fork north of present-day Abilene. The fort was never officially named, being called simply the "Post on the Clear Fork of the Brazos."

In the mid-1850s the Texas Legislature authorized two Indian reservations, one at Fort Belknap and one on the Clear Fork. The reservation on the Clear Fork became Camp Cooper and was Robert E. Lee's first command. Other notable figures of the Brazos were part of the region's history. Lawrence Sullivan Ross of Waco fought Indians in the area. Cynthia Ann Parker was abducted at Fort Parker, far to the east in the Navasota River watershed, but became the wife of Comanche chief Peta Nocona on the Brazos and mother of the most famous Comanche chief, Quanah Parker (Texas Beyond History 2009c).

The U.S. Army abandoned its activities in the region during the Civil War between 1861 and 1865. Texas did not have the capability to protect settlers, so local residents established Fort Hubbard on the Clear Fork in 1861, but it was a poor defense against the Indians. Indian attacks increased and many settlers left the area. In 1867 the federal government established Fort Griffin on the Clear Fork, near the old Camp Cooper. In 1871 William Tecumseh Sherman brought his Civil War tactics to the region, and by 1875 the Comanche and Kiowa people had been subdued (Texas Beyond History 2009b).

After the region was safe from Indian attack, hunters virtually eliminated the buffalo and ranchers replaced the buffalo with cattle. Railroads entered the region in the early 1880s and oil was discovered in the early 1900s. The region's economy has continued to be based largely on agriculture and petroleum.

The seventeen counties that are all or partially included in this part of the Brazos watershed currently have an estimated population of about 270,000 people. Abilene comprises about half of the region's population. After 1940 the Texas population shifted to cities. Excluding Abilene, the region's population has declined by 30 percent since 1940. Abilene's growth has not been comparable to that of the state as a whole. Its population increased by 187 percent from 1940 to 2008, but the state's population grew by 279 percent during that time (U.S. Census Bureau 2009a, 2009b).

From one perspective it appears that the four arms of the Brazos were not important in the modern development of the region due to their high mineral content and undependable flow. However, the major water supply for the region has been from small ponds and reservoirs built on creeks that are tributaries to the river. Although this water is not directly from the river, it is part of the Brazos system. Impounding and using that water removes much of it from the river's flow.

River Towns

Few towns were built on the Arms of God because the rivers vary so greatly between low flows and floods and because the Salt Fork is not useful for humans or livestock. However, four towns are notable exceptions: Griffin, Crystal Falls, Eliasville, and South Bend.

Griffin is the epitome of the transitory, rough-and-tumble settlement of the American frontier. The army did not consider Fort Griffin to be an important fort, and most of the buildings were flimsy greenwood shacks that leaked both wind and water and barely slowed visits from mice, skunks, and rattlesnakes. The administration building, bakery, and powder magazine were built of stone. The fort was situated above the river, which kept it safe from floods. The soldiers dug a forty-five-foot deep well, which still has water, plus a rock-lined cistern where they stored water hauled into the fort from Elm Creek, just west of the fort. The army built a sawmill on the creek they named Mill Creek, east of the fort. The park provides a nature trail along Mill Creek (Texas Beyond History 2009a, 2009c).

Because the Comanches and Kiowas were guerrilla fighters, the forts were not built as stockades with defensive walls. Rather, they provided places of logistical support for the army soldiers who chased and punished the Indians after their raids. Although the buffalo hunters who headquartered closer to the Clear Fork on the Flat below Fort Griffin were not military men, they were part of the government's strategy to defeat the Indians by eliminating their main resource.

The daily tedium of military life in the heat, cold, wind, and dust was accompanied by smelly horses, mules, and men. A commander said the post was "unfit for human habitation" (Texas Beyond History 2009a, 2009c). For more excitement, the place to be was on the river below the fort, in the community called the Flat or Griffin. This was the place to get drunk, diseased, stabbed, or shot, all in one night, possibly within an hour, probably in that order.

In the Flat the prostitutes' cribs were closest to the river, and these quarters would be first and most frequently flooded. The town's shifting population of sometimes one thousand people included such notables as Doc Holliday, Wyatt Earp, and John Wesley Hardin. Buffalo hunters brought in one thousand hides per week from 1874 until 1878, when they had decimated the buffalo herds. Cowboys replaced the buffalo hunters as the Western Trail, or Dodge City Trail, became the major route of cattle drives from Texas to the railheads in Kansas. Griffin was the last supply source on the northbound trail (Texas Beyond History 2009a, 2009c).

Crystal Falls is located about twenty-two straight-line miles southwest of Graham where Ranch Road 578 crosses the Clear Fork of the Brazos. The town was established in the 1870s partly to serve the buffalo hunting business. By 1892 there were several stores, a mill, and a gin. By 1900 Crystal Falls was the largest town in Stephens County. Discovery of oil brought the total population to twelve hundred in 1921, but the boom went bust and the people left (Davis 2009).

Eliasville is about thirteen straight-line miles southwest of Graham on the Clear Fork near the intersection of Ranch Roads 1974, 3109, and 701. By the late 1870s a flour mill and

The waters of the Clear Fork are often not very clear, so the soldiers at Fort Griffin used a tributary creek as their water supply.

a store were located there. Eliasville boomed in the 1918 oil rush but declined afterward (Hunt 2009a). Fewer than one hundred people live in the community today. The ruins of an old dam and mill provide a scenic place for fishing and playing in the river.

South Bend is about eight miles south-southwest of Graham on State Highway 67 near a huge southward bend of the Clear Fork of the Brazos. However, it was first named Arkansas. Like Crystal Falls and Eliasville, South Bend grew in the oil boom of the early twentieth century but then declined. One well drilled for oil instead produced hot mineral water, and the Stovall Hot Wells health resort operated there for a few years (Hunt 2009b).

Where to Experience the Land and the River

The land in this region is virtually all privately owned, so there is very little public access to the river and its tributaries. Road crossings do offer some access, but they are often brushy and may be used as illegal dumping grounds, so they are not attractive. There is a public park at the site of Fort Belknap, near Newcastle. Some of the military buildings have been restored, and the pleasant park gives a sense of the history of the region. However, it is not located on the river. Fort Griffin State Historic Site and Abilene State Park are good places to experience and learn about "Los Brazos de Dios."

FORT GRIFFIN STATE HISTORIC SITE: 1701 N. U.S. HWY 283, ALBANY, TEXAS 76430 325-762-3592

Fort Griffin State Historic Site north of Albany provides public access to the Clear Fork of the Brazos as well as a glimpse into the ecology and history of the Rolling Plains ecoregion. Fort Griffin State Historic Site has an interpretive visitor center with a small museum, an interpretive trail through the fort, and nature trails along the Clear Fork and Mill Creek, and a comfortable campground. The park is located fifteen miles north of Albany on U.S. Highway 283.

ABILENE STATE PARK: 150 PARK ROAD 32, TUSCOLA, TEXAS 79562 325-572-3204

Abilene State Park and the adjacent Lake Abilene are in the Rolling Plains ecoregion on the slope of the Callahan Divide at the southern edge of the Brazos watershed. To the south is the watershed of the Colorado River. Lake Abilene was one of the municipal water supply reservoirs built in 1921 under the provisions of the Conservation Amendment of 1917 that allowed cities to finance water supply reservoirs. The state park is below the dam, in the riparian area of Elm Creek. In the 1930s the Civilian Conservation Corps built a red sandstone water tower and swimming pool facility from rocks quarried locally.

This park is a good place to learn about the ecological effects of a small amount of water in a relatively dry land. The creek supports a rich bottomland riparian forest that attracted Tonkawa, Comanche, and Anglo settlers.

Old dams on the Clear Fork at Crystal Falls and Eliasville are reminders of efforts to establish communities in places that were not very hospitable to permanent human settlement.

John Graves's Dammed River

After the dam was finished at Possum Kingdom near the beginning of the war, it began to filter out the West Texas drainings, and that piece of the Brazos ran clear for more of the year than it had before, and the old head rises no longer roared down, and the spring floods were gentler and the quicksands less quick. But it was there still, touchable in a way that other things of childhood were not.

John Graves, Goodbye to a River (1960)

IT IS WITH BOTH HONOR and irony that we call this section of the Brazos "John Graves's Dammed River." Graves wrote Goodbye to a River in gentle protest to the Brazos River Authority's plan to build six additional reservoirs above Lake Whitney, a plan that would have turned the middle Brazos into a string of lakes 250 miles long. Only one of the dams that Graves protested was ultimately built, but many that he did not anticipate were built on Brazos tributaries. There are now forty-four major dams in the Brazos system, and several more are under consideration.

We have defined this section of the river by the influence of dams, beginning with Possum Kingdom on the mainstem of the Brazos and extending to the mouth of the Navasota River, which is dammed in its upper reaches. This is a much larger area than Graves wrote about, but it seems fitting to include all of the dammed rivers in the Brazos system because they ultimately affect the characteristics and behavior of the rest of the river, as we will discuss in the next chapter.

MORRIS SHEPHARD (POSSUM KINGDOM) DAM NEAR MINERAL WELLS, TEXAS

Completed in 1941, the same year Woody Guthrie extolled the dams on the Columbia River, Morris Sheppard Dam was the first such structure on the mainstem of the Brazos.

John Graves's Dammed River

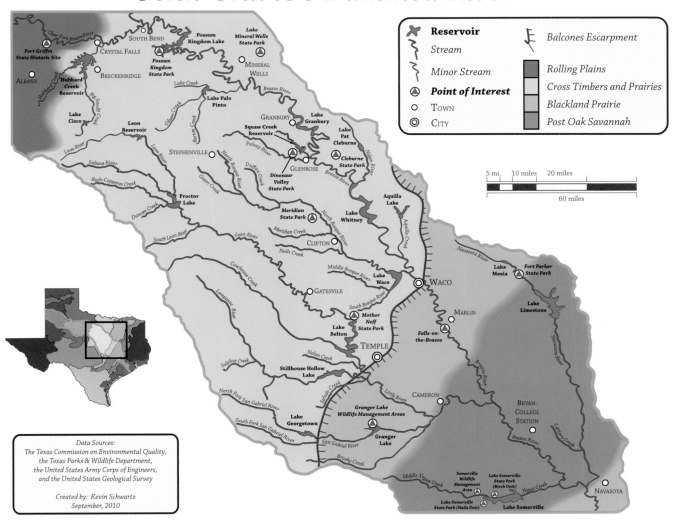

Legend

Reservoir
Stream
Minor Stream
Point of Interest
○ TOWN
◎ CITY

Balcones Escarpment
Rolling Plains
Cross Timbers and Prairies
Blackland Prairie
Post Oak Savannah

5 mi. 10 miles 20 miles
60 miles

Data Sources:
The Texas Commission on Environmental Quality,
the Texas Parks & Wildlife Department,
the United States Army Corps of Engineers,
and the United States Geological Survey

Created by: Kevin Schwartz
September, 2010

Morris Sheppard Dam stands about 190 feet above the riverbed. The trip that John Graves chronicled in Goodbye to a River *began here, just below the dam.*

Water of the Dammed River

CLIMATE

Rainfall in this section of the river varies from a yearly average of 31.35 inches upstream at Graham to 33.34 inches downstream at Waco. The region has two rainy seasons, as illustrated by the charts in the rainfall sidebar. The spring rains result from moist air from the Gulf of Mexico that is pushed aloft by cool dry air from the north. The late summer and fall rains come from hurricanes that sweep across the land from the Gulf and even from the Pacific Ocean. Averages shown in the charts are deceptive, however. What is not evident is that a month's total precipitation may

fall in one or two storms, and thus the river will carry large amounts of water for short periods of time.

FLOW

The natural timing and amount of rainfall are the primary determinants of the Brazos River's flow. Before the dams were built on the mainstem the highest flow recorded at Waco was 246,000 cubic feet per second (cfs) on September 27, 1936. In the ten years before Lake Whitney was completed upstream of Waco the average monthly flow varied from 195.9 cfs in October 1940 to 22,470 cfs in April 1942.

The major tributaries to this section of the Brazos are even "flashier" (see the sidebar on flows of major

We can better understand the effects of uneven rainfall distribution and high temperatures if we compare the Brazos to the river Thames in England. The Thames is an important river in history, being a major transportation artery for the Industrial Revolution and the focus of transportation and power for the huge and wealthy British Empire, including the American colonies. To be so important in history it would seem that the river Thames must be a large river, fed by the high rainfall of England, but that is not so. The Thames is an impressive and pleasant river, but it is not very big and the rainfall that supplies it is surprisingly low. The lower Thames is an estuary and receives substantial tidal flow, but even the nontidal parts of the Thames have been important in English history. The river Thames is much shorter than the Brazos, only 215 miles compared to more than 1,200 river miles of the Brazos. Its average flow is about one-third that of the Brazos, but its catchment area is only about one-tenth of the Brazos watershed, implying that the smaller Thames catchment area must receive much more rainfall than that of the Brazos. Contrary to our stereotype, near

London the average annual rainfall is only about twenty-two inches, substantially less than what falls in this section of the Brazos. How can such a small watershed with relatively low rainfall produce an important river like the Thames?

There are two parts to the answer. The first is that the rainfall that feeds the Thames is fairly evenly distributed through the year unlike precipitation in the Brazos watershed, which has two periods of higher rainfall separated by longer periods of drier weather. This dependable source of moisture in the Thames watershed supports lush vegetation, which over millennia has developed deep soils that absorb water and release it to the river slowly as baseflow. Thus, the Thames basin is able to retain more of its rainfall than the Brazos can.

Heat is the other part of the answer. The average high temperature in both Graham and Waco is between 90 and 96 degrees Fahrenheit in July and August but only 70 to 74 degrees Fahrenheit in London. High heat increases water loss in the Brazos watershed because hot air can absorb more water than cool air.

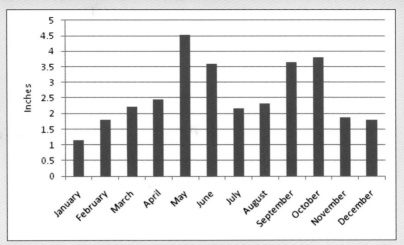

Average monthly rainfall at Graham, Texas, 1971–2000.
Source: NOWData – NOAA Online Weather Data. http://nowdata.rcc-acis.org/FWD/pubACIS_results.

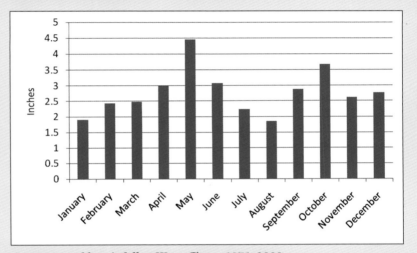

Average monthly rainfall at Waco, Texas, 1971–2000.
Source: NOWData – NOAA Online Weather Data. http://nowdata.rcc-acis.org/FWD/pubACIS_results.

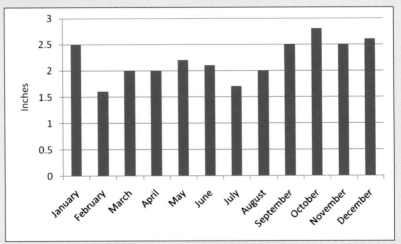

Average monthly rainfall at Bracknell, United Kingdom, twenty-eight miles west of London, 1971–2000.
Source: Met Office. http://www.metoffice.gov.uk/climate/uk/so/print.html

tributaries). The North Bosque has a recorded high flow that is 15,068 times greater than its lowest annual flow. These normally small tributaries drain a landscape of thin soils that do not absorb much water, so the torrential rains of the region produce major runoff. The first flood I saw was in 1950 at Clifton on the North Bosque. I stood on the bridge, fascinated by the water that looked like chocolate milk, but it was roaring and tossing entire trees.

Although the Little Brazos River is not included in the table showing flows of major tributaries, it is a stream that drains the Hearne area and joins the Brazos just west of Bryan–College Station, but its flow is not monitored by the U.S. Geological Survey. It is interesting, however, because its 72-mile length roughly parallels the Brazos, making it what hydrologists call a yazoo stream, named after the Yazoo River, which runs through the state of Mississippi and parallels the Mississippi River for about 175 miles.

The highly variable flow of the Brazos mainstem and its tributaries makes it obvious that floods are large and common on these rivers. Floods on the lower Brazos in 1833 and 1842 were disruptive to the agricultural economy of that part of the state

during its early years. In 1913 people died and huge amounts of damage occurred as the Brazos flooded eastern Waco and the Brazos and Colorado rivers merged, inundating more than 3,000 square miles (Burnett 2008). In September 1921, a tropical storm dropped 38 inches of rain at Thrall, causing the Little River to run at 647,000 cfs, greater than the average flow of the Mississippi. The topography of the region is relatively flat, so the flood waters spread widely, inundating many communities, including Cameron, and causing several deaths (Scarbrough 2005; Burnett 2008).

Private landowners built levees to try to prevent flooding, but these were uncoordinated and actually increased flooding. Drainage districts were established and their management tried to control floods with coordinated levees but did not have the resources to build enough of them. By the early part of the twentieth century public officials were calling for dams in the middle Brazos to prevent flooding along the lower river (Committee on Flood Control 1919).

The dams are discussed in detail later in this chapter. However, since this section focuses on the river's flow, we should note that the dams have substantially changed its characteristics so the flow no longer serves some of its important ecological functions. Specifically, before the dams retained the river's flow, typical May and June flows would be in the range of 1,200 cfs. Now they are about 450

cfs. Likewise, typical flows in September and October before the dams were about 400 cfs. Now they are about 100 cfs. Perhaps most important, there is little variation between July and December. In addition to reducing high flows, the dams have decreased the frequency of low flows. These changes in flow can have important effects on both the biotic and physical characteristics of the river (Trungale Engineering & Science 2007).

WATER QUALITY

Before thinking about dams and reservoirs, however, we must understand the characteristics of the water that is held in those reservoirs. The Texas Commission on Environmental Quality maintains a comprehensive water quality program for the state's rivers called the Clean Rivers Program, and it works closely with river authorities and other entities to implement the program. The Clean Rivers Program requires a summary assessment report of water quality conditions every five years. As seen in the sidebar about the 2007 Clean Rivers Program report (p. 85), nutrients and bacteria are the main pollutants of this part of the Brazos.

This section of the Brazos receives runoff from all or parts of twenty-three counties, which together have a total population of almost 1.4 million (U.S. Census Bureau 2009). Wastewater, including many on-site systems, is a source of nutrients and bacteria, as are various agricultural operations and wildlife. All developed areas produce some level of nonpoint source pollution that results from stormwater runoff. In this context it is actually a little surprising that water quality problems are not more severe.

It is notable that the mainstem of the Brazos from Waco to the mouth of the Navasota exceeds bacterial standards for contact recreation, that is, swimming. In fact, few people swim in this part of the Brazos. However, as we emphasize later, the Brazos below Waco is an important remnant of wilderness in our highly developed state. People do use the river for river trips, and Hidalgo Falls is a popular place to run rapids. I strongly encourage increased recreational use of the Brazos; therefore, we should take seriously the issues affecting its water quality.

The Bosque, where I first experienced the delight of a river, has been the focus of conflict between the City of Waco and the dairy industry in the Stephenville area. Waco charged that nutrients from the dairies polluted Lake Waco, the city's source of drinking water, while the dairy industry maintained that it met state requirements for its waste. This complex conflict continues, but the number of dairy cattle has decreased in recent years.

Land of the Dammed River

GEOLOGY AND TOPOGRAPHY

Is there something special about this part of the Brazos watershed that would entice people to build so many dams here? Yes, it's the geology. The best place to build a dam is in a canyon with steep walls of rock that do not absorb much water. Sandstone is not very good, as was learned at Glen Canyon in Arizona, where the sandstone absorbed so much water from Lake Powell that the lake took twenty years to fill and springs arose on the adjacent Navajo lands where none had existed before. A narrow limestone canyon is ideal. A canyon requires only enough dam to plug the canyon, and it provides a natural basin for the lake. Morris Sheppard Dam, which forms Possum Kingdom Lake, is a good example, at least for Texas, where we don't have many really deep canyons. DeCordova Bend Dam, which forms Lake Granbury, is not in a canyon, but Whitney Dam is in a steep, narrow valley.

A limestone canyon was the ideal location for Morris Sheppard Dam, which forms Possum Kingdom Lake.

Stream and measurement location	Average annual flow in cfs (time period)	Range of annual flows in cfs (year)	All-time peak flow in cfs (date)
Paluxy River at Glen Rose	96.1 (1982–2007)	6.24 (1984) to 361.3 (1992)	59,000 (1908)
Nolan River at Blum	NA	NA	79,600 (May 17, 1989)
Aquilla Creek near Aquilla	121.8 (1983–2001)	2.24 (1984) to 395.5 (1992)	74,200 (1936)
North Bosque River at Valley Mills	334.9 (1959–2007)	14.6 (1984) to 2,706 (2007)	220,000 (December 21, 1991)
Little River near Cameron	1,798.1 (1954–2007)	173.5 (1956) to 7,759 (1992)	647,000 (September 10, 1921)
Leon River at Gatesville	337.4 (1954–2007)	6.22 (1978) to 1,349 (1997)	70,000 (1908)
Sabana River near DeLeon	270.4 (1960–2007)	0.21 (1960) to 1,883 (1994)	15,400 (June 14, 1989)
Lampasas River at Youngsport	272.8 (1924–2007)	19.5 (1951) to 752.2 (1944)	87,900 (May 17, 1965)
San Gabriel at Laneport	260.4 (1980–2007)	18.4 (2000) to 1,015 (1992)	ca. 40,000 (September 1921)
Yegua Creek at Somerville	284.2 (1924–1967, prior to dam)	0.46 (1925) to 929.1 (1941)	56,800 (July 1, 1940)
Navasota River near Bryan	634.5 (1997–2007)	102.2 (2006) to 1,355 (2007)	66,600 (December 23, 1991)

Source: U.S. Geological Survey Surface-Water Annual Statistics for the Nation 2008.

SOILS, VEGETATION, AND LAND USE

The upper part of this reach of the Brazos runs through the Cross Timbers and Prairies ecoregion, with its grasslands and scattered post oak, blackjack oak, and increasing stands of juniper and mesquite on the thin, sandy loam upland soils. Elm, pecan, and hackberry grow in the deeper valley soils. The land is mostly used for grazing, but valley soils support vari-

ous row crops. Oil and gas wells are common. The land is gently rolling to hilly, with elevations ranging from about 1,000 feet at the upper end of Possum Kingdom Lake to 390 feet near the Lake Brazos Dam at Waco, giving the river an overall slope (gradient) of 5 feet per mile.

From just below Waco to the confluence with the Navasota River the Brazos runs through the gently sloping Blackland Prairie and Post Oak

Savannah, where the overall gradient is about two feet per mile. These ecoregions consist of deep soils deposited by seas and rivers during the past several million years. In its natural state, the land was native grassland with thick riparian forests adjacent to the river.

The Six Dam Plan did not materialize to provide irrigation water below Waco. However, as discussed in chapter 3, the Ogallala Aquifer,

River segment	Water quality concerns	Causes
Upper Brazos watershed (from confluence of Salt and Double Mountain forks to Lake Whitney)	Increasing trend of nutrient enrichment in Possum Kingdom Lake, Lake Granbury, and Lake Whitney	Municipal discharges, malfunctioning on-site wastewater disposal, agricultural runoff
Aquilla Creek	Nutrients in Lake Aquilla	Sources not determined but may include wastewater discharges and nonpoint source runoff
Bosque River	Elevated levels of phosphorus and bacteria; nutrient enrichment and algal growth in Lake Waco	No causes stated in the report
Leon River	Elevated bacteria and nutrient levels; increasing nitrate-nitrite levels in Lake Belton	Municipal discharges, malfunctioning on-site sewage disposal systems, residential and agricultural runoff, and concentrated animal feeding operations
Lampasas River	Previous problems of bacteria and dissolved oxygen deficiencies appear to have been resolved.	
Little River	Good water quality overall; some nutrient concerns in Lake Granger, Willis Creek, Brushy Creek, and Mankins Branch	None stated
Central Brazos (from Lake Brazos Dam at Waco to the mouth of the Navasota River)	Bacterial levels exceed standard for contact recreation in the Brazos River, Thompsons Creek, Campbells Creek, Mud Creek, Pin Oak Creek, Spring Creek, Tehuacana Creek, and Big Creek; nutrient enrichment in Thompsons Creek	None stated
Navasota River	Generally good water quality, but certain segments exceed standards for fecal coliform and dissolved oxygen.	Sluggish stream flow, warm temperatures, abundance of organic matter
Yegua Creek	Lake Somerville does not meet standards for general use due to low and high pH levels. Davidson Creek is impaired by elevated bacterial concentrations.	Algal photosynthesis and respiration are possible causes of pH fluctuations.

which irrigates the highly productive agriculture of the High Plains, will not last forever, so sometime in the future people may look again to the Brazos to irrigate food crops, but it is likely that the river's salt would ultimately ruin the soil.

Ecology of the Dammed River

ECOLOGICAL EFFECTS OF DAMS

Consider the three perspectives on material and energy flows within rivers discussed in chapter 2: the river continuum concept, the flood pulse concept, and local riparian production. Dams have profound effects on all of these processes of river ecology.

The river continuum concept describes how each part of a river is influenced by the river processes upstream. A dam is literally a "stopper" for material and energy flows down the river. The flowing water slows and drops its sediment load, mostly in the upper parts of the reservoirs. The water that flows beyond the dam

The central part of the Brazos River valley is extensively used for economic production ranging from agriculture to mining and even nuclear power generation.

Farming the Brazos Bottoms

The Six Dam Plan that the Brazos River Authority put forward in the 1950s was partly supported by the prospect of providing irrigation water for almost two million acres of alluvial soil below Waco (Lockwood & Andrews 1955). Given that those dams were not built, farmers in the region continue to depend on rainfall and groundwater for irrigation because the variable flow of the Brazos is not dependable enough for irrigation. However, we visited a turfgrass farm that pumps water from the river into small reservoirs that supply canals from which it irrigates the grass. This is an expensive operation, but in today's strange agricultural economics it is more profitable to grow turfgrass than food, so the cost is justified.

Even though agriculture faces great economic difficulties and is often criticized, good farmers and ranchers use their intelligence and integrity to persevere in their passion for keeping their land productive. Penne and Mark Jackson at Waco and Connie and John Giesenschlag at Snook spent entire days helping us understand the agriculture of the middle Brazos. What we learned is that if you are very smart, work hard and long, carry huge debt, and don't expect any of the privileges corporate executives take for granted, you might be able to make a relatively good living most years, but not all. On the other hand, you will spend most of your time on your land, pretty much making your own decisions—within the context set by a host of laws, government programs, and the bank.

These are truly the American family farmers we hear so much about. They are successful because they adapt and diversify, just like a natural ecosystem. Penne and Mark operate a dairy on the banks of the Brazos. In a complex flow of energy and materials they not only produce dairy products but also provide straw and manure for mushroom production and ultimately use much of that material to restore agricultural production to abandoned quarry pits in the Brazos floodplain. Over decades John's farming has evolved from row crop production that required fifteen passes per crop over the fields with a tractor to newer methods that require only two to three passes, consequently using less fuel and fewer chemicals. He is now producing watermelons and thinks that a farmer could actually make a good living producing vegetables for sale at farmers' markets.

is usually clear because it has very little sediment. That means that it has powerful erosive capacity downstream of the dam. In terms of the river continuum concept, the river essentially starts over below a dam.

The flood pulse concept hypothesizes that a substantial part of the ecological production of a river comes from material and energy washed in by floods from the floodplain. Since a major purpose of the dams in the Brazos system is to reduce floods, it is obvious that this form of material/energy transfer is greatly inhibited.

The local riparian production concept asserts that an important part of the material and energy for biological production in the river comes from vegetation in the adjacent riparian corridor. However, the surface area of a reservoir is much greater than the surface area of the river it replaced. Consequently, the influence of the adjacent vegetation is greatly reduced. Also, due to the topography of the reservoir basin and frequent fluctuations of the water level, it is common for there to be relatively little vegetation along the shore.

Thus, the ecological productivity of much of the Brazos system is probably affected by these disruptions in the flow of material and energy. We do not know how much the productivity was affected because we do not have relevant data for the time before the dams were built.

Dams have profound and specific effects on rivers. The reservoirs

Corn growing in a restored riverside quarry illustrates the fact that intelligent use of riparian land can have both economic and ecological benefits.

with mud, as happened at the original Lake Waco on the Bosque River. Dams diminish downstream flow, a circumstance we experienced when our kayaks dragged bottom on the Brazos below Possum Kingdom Dam and Lake Whitney. In a hot climate like that of Texas evaporation losses are substantial. Actual evaporation is difficult to measure, but hydrologists estimated that evaporation in the Brazos basin is more than nine hundred thousand acre-feet per year (Center for Research in Water Resources 1997). This is a notable portion of the river's total average flow of about six million acre-feet. Dams either eliminate seasonal flows or change their sequence and magnitude so much that natural cycles of high and low flows no longer function. Water moves slowly through a reservoir, where nutrients and heavy metals can concentrate, oxygen is consumed, and methane gas is produced. Aquatic species composition shifts from riverine to still-water types that are sometimes not native, species richness may decline to as low as 10 percent of the riverine system, and natural essential migrations of fish and other animals are disrupted (Cooke et al. 1986; Jobin 1998; World Commission on Dams 2000; Wetzel 2001).

However, one of the major objections to dams may not apply to the Brazos. Dams pose a major threat to estuaries, those incredibly important habitats that often develop where rivers meet the sea. Estuaries are the

formed by dams inundate huge amounts of rich bottomland and often flood towns and displace people as well. Linda Scarbrough (2005) writes poignantly of the Czech and Moravian farmers displaced by Granger Lake on the San Gabriel River. Dams trap the sediment load of rivers and can ultimately fill reservoir basins

The rocky shoreline of Possum Kingdom Lake is dramatic and offers good access for swimming, but its biological productivity is low compared to that of a natural riverbank.

verted to a new mouth where it empties directly into the Gulf without an enclosed bay that could form a large estuary such as Galveston Bay or San Antonio Bay. However, the lower few miles of the river do contain a mixture of fresh and salt water and serve as a minor estuary.

Since this part of the Brazos system is so greatly modified, it seems fitting to focus on two species that flourish here because of the modifications. The striped bass and its nemesis, golden algae, are the species that seem to best represent the life of this purposely and drastically altered section of the river.

STRIPED BASS (*MORONE SAXATILIS*)

The striped bass is a coastal species, native to shores, bays, and estuaries along the Atlantic coast from New Brunswick to Florida and as far west as Louisiana. It is an anadromous species, like salmon, that spawns in fresh running water but spends its adult life in brackish or salt water. Stripers in salt water may reach 100 pounds, but the largest caught in fresh water so far has been 67.5 pounds. Stripers have been an important commercial and sport fish on the Atlantic coast since early colonial times. Specialized boats are now used for offshore striper fishing. Obviously, the opportunity to catch such a fish in a lake near home, perhaps even from the shore without an expensive boat, excites many Texas anglers. Tens of millions of

nurseries of important marine species and are some of the most productive ecosystems on Earth. Their unique character and capabilities depend on certain amounts and timing of freshwater inflow, which also bring nutrients to the system. Dams disrupt and reduce this freshwater inflow,

posing incalculable risk to marine systems. But the Brazos doesn't flow into a large bay estuary. The Brazos, Colorado, and Rio Grande filled in their bays long ago with their heavy silt loads, so their estuarine functions are limited. Also, as we will see in the next chapter, the river's flow was di-

Lake Whitney was built in a particularly scenic section of the Brazos where limestone cliffs towered above the river. Although dams severely impact natural river systems, the reservoirs (lakes) they form are often beautiful and very popular.

dollars are spent each year on striper fishing in Texas. Lake Texoma alone, on the Red River, generates about twenty million dollars per year from striped bass fishing. The record striped bass in Texas, caught in the Brazos just below the dam that forms Possum Kingdom Lake, weighed 53 pounds and measured 48 inches long (Texas Parks and Wildlife Department 2008).

How did a coastal fish come to be so popular in Texas lakes? In the mid-1950s fisheries biologists in South Carolina inadvertently trapped adult stripers in a reservoir and found they could survive in freshwater habitats. The fish were first released in Texas in 1967. Striped bass now naturally spawn in Lake Texoma, on the Red River, but not in any of the Brazos reservoirs. Consequently, the Texas Parks and Wildlife Department annually stocks reservoirs with striped bass, as well as other game fish. In 1993 the agency put 5,115,522 striped bass fry in Possum Kingdom Lake alone (Texas Parks and Wildlife Department 2008). Stocking rates for the Brazos reservoirs in 2009 were: Possum Kingdom 269,156; Granbury 44,864; and Whitney 543,856 (Texas Parks and Wildlife Department 2010a).

Fishing is an important form of

recreation in Texas. In its 2006 report the U.S. Fish and Wildlife Service noted that more than 2.5 million Texans participated in fishing the previous year, for a total of more than 41 million fishing days. Those anglers spent more than $3.2 billion (U.S. Department of the Interior et al. 2006). Thus, we can understand the concern about another introduced river species: golden algae.

GOLDEN ALGAE (*PRYMNESIUM PARVUM*)

Biologists have known golden algae as an estuary species in Europe and the Middle East for decades. They found the algae in the United States only in the mid-1980s. These algae produce their own food via photosynthesis, but they are also predatory and release a toxin to immobilize their microscopic prey (Skovgaard and Hansen 2003; Morgan 2006). Unfortunately, the toxin also affects fishes' cell tissue, including their gills, causing hemorrhaging and ultimately death (Texas Parks and Wildlife Department 2010b). Possum Kingdom Lake was hit hard by golden algae in 2001 and 2003, losing hundreds of thousands of striped bass and other fish. Even the fish hatchery there was infected and lost its crop.

At this point the only known method of reducing the effects of golden algae is to increase the water's acidity by applying ammonia to the water. This is possible in the ponds of a fish hatchery but not possible in the huge volume of water in a reservoir. Thus, the popular and profitable fish populations of Possum Kingdom, Granbury, and Whitney lakes are continually threatened by attacks of golden algae, and there is little defense against the threat.

People and the Dammed River

PREHISTORY

Central Texas, including the Brazos River valley, apparently has been hospitable to people for millennia. The archeologist Michael Collins has written that "central Texas was one of those places in the world where the labors and limitations of food production could be looked upon with disdain. What then are the ingredients of that technology and the characteristics of that resource base? The axiom that specialization is the path to extinction seems to be borne out by its corollary, namely eleven thousand years of successful, generalized exploitation of a diverse resource base in central Texas" (Collins 2004, 124).

The Brazos River basin continues to help us understand the deep history of the New World. As discussed in chapter 3, the Clovis people have generally been considered the earliest people in the New World, and they were named for the site where their artifacts were first found, in Blackwater Draw near Clovis, New Mexico. The Brazos and its tributaries are a major center of continuing research about the Clovis people and the cultures that followed them. The Lubbock Lake site discussed in chapter 3 illustrates how even a small amount of water sustained human life for thousands of years. Farther down the Brazos, the Horn Shelter No. 2 in Bosque County between Lake Whitney and Waco contains archeological materials that span the time from 10,000 BP to the 1930s. The Bosque Museum in Clifton has an exhibit that interprets the Horn Shelter site. The Bell County Museum in Belton interprets the Gault site from a Brazos tributary. The Wilson-Leonard site in southern Williamson County has one of the most complete cultural sequences in North America, ranging from 12,000 to about 800 BP.

HISTORIC INDIANS

Simple generalizations about historic Indians in the region are difficult because there was a continual ebb and flow of people moving in, displacing others, and then being displaced themselves. None had permanent settlements. A prominent scholar of Texas Indians, David La Vere (2004, 115) has written that "by the mid-seventeenth century Central Texas was filled with Indian peoples, some native to the area, others hailing from outside regions. Most, but not all, were hunter-gathers similar to the Coahuiltecans, living off buffalo, deer, and other animals and plants."

We know that the Apaches moved into the region sometime shortly before the Spanish explorers arrived. The profusion of different people was made even more confusing by shifting allegiances. La Vere (2004, 115) writes that "ethnogenesis took place constantly in Texas as Indian peoples made and remade themselves. Bands split, rejoined, were absorbed by other cultures, or were pulled into Spanish missions where they created kinship with other groups." In 1684 a trade fair on the upper Colorado River south of the Brazos basin was attended by fifty-seven different Indian nations (La Vere 2004). However, five groups became predominant in the Brazos area: Apaches, Wichitas, Caddos, Comanches, and Tonkawa. The Apaches were relative newcomers from Canada. The Wichitas possibly evolved from the Henrietta focus of the upper Panhandle area that disappeared after 1500 (Campbell 2003). The Caddos were an eastern group that had been pushed westward. The Comanches were former hunter-gatherers who evolved from the Shoshoni and turned into a new and powerful people when they got horses. The Tonkawa were the result of "ethnogenesis" as smaller tribes coalesced in response to the powerful newcomers (La Vere 2004, 115).

Most of the historic Indians except the Comanches combined hunting and gathering with small-scale agriculture. Thus, the streams, riparian areas, and floodplains of the Brazos basin would have been important because they provided habitats for the animals and plants the Indians used and had suitable soil for agriculture. Certainly, the streams would have been sources of drinking water. There is no indication that the people of the central Brazos basin used rivers for irrigation or transportation.

In 1846 the German geologist Dr. Ferdinand Roemer traveled about sixty miles from Torrey's Trading Post on Tehuacana Creek south of the Waco village and visited a Caddo village. This would place the Caddo village somewhere near today's Glen Rose. Roemer's description of the

A tributary stream enters the Brazos, linking its riparian zone to that of the Brazos, thus forming part of a network of interconnected habitats that maintain biodiversity and have provided rich hunting grounds for people for thousands of years.

area, written after two days' travel, is consistent with that location:

> After a ride of about thirty miles we saw, toward sundown, from the top of a hill, the destination of our excursion—the Caddoe Indian village lying before us. A more suitable and pleasant place could not have been selected by the red sons of the wilderness. The village lies in the center of a plain two miles long which on the one side is bordered by the wooded banks of the Brazos and on the other by steep precipitous hills. A beautiful clear creek flows diagonally through this plain on a smooth bed of limestone and along its banks are several large live oaks. The huts of the Indians stood on both sides of this creek in picturesque disorder and near each was a cornfield. Between the hills from which we looked down and the village proper, about a thousand head of horses were grazing in the plain, among which a number of naked, long haired Indian boys rode back and forth yelling. . . .
>
> Just after sunrise on the following morning we took a walk through the village. The home of every family consisted of several huts of diverse form. There is always a large conical shaped hut present, about fifteen feet high, which is enclosed on all sides except for a small opening at the bottom. It is thatched with long grass and therefore at a distance resembles a haystack of medium size. This hut is used in cold and wet weather. Near it are several other huts, open on the sides, which really are only grass-covered sheds, resting on four uprights under which at a distance of about two feet from the ground, a horizontal latticework platform, woven from strong twigs, is fastened. On this wicker platform men and women squat during the hot hours of the day. The roof shelters them from the hot sun's rays and at the same time the air can circulate freely on all sides, even from the bottom. Finally there was a third kind of hut used for storing provisions which was nothing but an oven-like container also covered with grass, resting on four, high posts. (Roemer 1935, 200–201)

Roemer described an almost idyllic setting, but from the long view of history things were not going well. Randolph Campbell (2003, 200) writes that "conditions on the western Texas frontier during the late 1840s and 1850s amounted to a recipe for disaster, especially for the fragmented peoples lumped under the heading 'Indians.' . . . Pressured by settlers from the east, the agriculturalists were unable to stay in one place and grow crops and the hunters saw buffalo herds begin to decline in size."

After 1854 the people Roemer visited joined other groups on the Indian Reserve upstream on the Brazos (La Vere 2004). However, in 1859 a commission authorized by Texas governor Hardin Runnels stated, "We believe it impracticable if not impossible, for tribes of American Indians, scarcely advanced one step in civilization, cooped up on a small reservation and surrounded by white settlers, to live in harmony for any length of time" (quoted in Campbell 2003, 203). And the people were herded across the Red River into Indian Territory.

EUROPEANS

The Spanish explorers arrived in what is now Mexico in the early 1500s, and a century later they began following the Rio Grande northward from El Paso seeking minerals and converts to Christianity. They did not pay much attention to Texas because it did not have the kind of mineral wealth they sought and the Rio Grande pueblos provided sufficient opportunity for their missionary activities. In 1685 the French explorer René-Robert Cavelier, Sieur de La Salle, established a colony on Matagorda Bay, thinking it was one of the mouths of the Mississippi. The colony failed, with almost everyone dying of disease or starvation or being killed by the Karankawa, but the presence of the French spurred the Spaniards to try to protect the territory they believed was theirs. Thus, from 1690 through the early 1720s they established missions and forts on the Neches, Sabine, and Red rivers in what is now East Texas and western Louisiana. These missions and forts were hundreds of miles from their supply points in northern Mexico and ultimately failed. In their focus on the French threat in the east, the Spaniards ignored the central part of their colony, including the Brazos (Foster 1995; Campbell 2003).

In the 1740s three Indian groups living on a middle Brazos tributary, the San Gabriel, requested that the Spaniards provide missions for them. These groups reflected the disruption of native life during the time. The Cocos and Deadoses had migrated from the coast. The other group was an amalgam of smaller tribes called the Tonkawa (Campbell 2003). It is likely that these people hoped the Spaniards would protect them from the Apaches and Comanches.

The Spaniards built a mission for each of the groups between 1746 and 1749. In 1751 they built a fort to protect all three missions. Presidio San Francisco Xavier de Gigedo, near present-day Rockdale, was supplied with fifty soldiers under the command of Captain Felipe de Rábago y Terán. Upon arrival Captain Rábago reported that the missions were failures and should be abandoned. However, his own behavior of seducing a woman and then ordering the imprisonment and torture of the cuckolded husband, and perhaps even the murder of the husband and a priest, set a model for the rest of the soldiers that led to the short-term excommunication of Captain Rábago and all of his soldiers. The Indians were dismayed and left. The Spaniards moved the missions to the San Marcos River in 1755 (Campbell 2003).

The Spaniards had little influence on the Brazos and abandoned it. They took over Louisiana in 1769, which may have made Texas even less attractive. General Commandant Teodoro de Croix of the Interior Provinces of New Spain expressed his disdain of Texas in the late 1770s: "A villa without order, two presidios, seven missions, and an errant population of scarcely 4,000 persons of both sexes and all ages that occupies an immense desert country, stretching from the abandoned presidio of Los Adaes to San Antonio . . . [that] does not deserve the name of the province of Texas . . . nor the concern entailed in its preservation" (quoted in Campbell 2003, 75).

A series of revolts against the Spaniards beginning in the early 1800s ultimately displaced them in 1821, and the Brazos became part of the Republic of Mexico. The Spaniards failed to "civilize" the Indians in Spanish terms, but they greatly added to a disruptive period of time for the native people. The Spanish incursion did leave three lasting legacies related to rivers. First, Texas water law that gives the public ownership of surface water was originally instituted by the Spaniards. Second, irrigation dams and canals that the Spaniards built on the San Antonio River in the early 1700s continue to function today. And third, all of the major rivers in Texas have Spanish-origin names except the Red River (Campbell 2003).

The Republic of Mexico was no more able to defend and colonize its northern territory than was Spain, but it took a different approach that ultimately brought people to this section of the Brazos and its tributaries. Whereas Spain vigorously defended its territory against incursions from the United States, Mexico chose to try to use Anglo American colonization as a means of populating the region and thus defending its territory. As we will see in the next chapter, the Brazos River was an important part of this colonization process. The naïve strategy of using American settlers to defend the territory from other Americans went awry, and Texas declared its independence in 1836, opening the doors wide to emigration from the United States.

The Americans were aggressive and quickly spread from the concentration of settlements along the coast. In 1842 and 1843 the Torrey brothers established trading houses on the Bosque and Navasota rivers, Tehuacana Creek in McLennan County, and the Falls of the Brazos near present-day Marlin (Armbruster 2005). On his 1846 visit to the region the German geologist Dr. Ferdinand Roemer traveled with John Torrey and wrote about the Torrey Trading House on Tehuacana Creek:

On rounding a corner of the forest, we suddenly saw the trading post before us. It lay on a hill covered with oak trees, two miles distant from the Brazos, above the broad forested bottom of

Tohawacony Creek. The entire layout consisted of from six to seven log houses built in the simple customary style out of rough, unhewn logs. These houses were not surrounded by palisades, as are those of the trading companies on the Missouri, neither do they contain any other protective enclosures. The safety of this trading post against possible Indian attacks is founded rather on its usefulness, in fact its necessity to the Indians.

The largest of the log houses contained the pelts received in trade from the Indians. Buffalo robes or buffalo rugs and the hides of the common American deer (Cerevus Virginianus L.) [sic] formed by far the greatest number of hides. Some of the buffalo skins are brought to the trading post entirely raw, some are tanned inside only, and very often they are painted more or less artistically. Their value depends upon the size, the uniformity of the fur, and also upon the artistic paintings on the inside. The hides of average quality were sold in Houston for three dollars, and the fancy ones for from eight to ten dollars. Most of them are shipped to the Northern States, also to Canada, where they are used for covers in sleds or for wagon seats. Leather is not made out of the skins since it is too porous and not compact enough. (Roemer 1935, 191–192)

Commerce was bringing the Brazos to the larger world, and within a few decades railroads, farms, towns, cities, and dams would change it drastically.

FORTS, TOWNS, AND CITIES ON THE BRAZOS

As Anglo Americans settled Texas in the 1800s they sought places that could provide water, their most crucial resource. They needed drinking water for themselves and their animals. They also needed water to power their gristmills and sawmills.

Cowboy's Brazos River Song

River crossings were difficult and dangerous for the young men who drove cattle from Texas to the railroads in Kansas after the Civil War. The Chisholm Trail crossed the Brazos at Waco. The "Brazos River Song," also known as the "Texas River Song," perhaps tells us that the cowboys were thinking more about women than the dangers of rivers.

Verse 1:
We crossed the wild Pecos; We forded the Nueces
We swum the Guadalupe; We followed the Brazos
Red River runs rusty; The Wichita's clear
Down by the Brazos; I courted my dear.

Verse 2:
The fair Angelina is glossy and glidy
The crooked Colorado is twisty and windy
The slow San Antonio courses the plain
I'll never walk by the Brazos again.

Verse 3:
She hugged me, she kissed me; She called me her dandy
The Trinity's muddy, the Brazos quicksandy
She hugged me, she kissed me; She called me her own
Down by the Brazos she left me alone.

Verse 4:
The girls of Little River; they're plump and they're pretty
The Sabine and Sulfur have beauties a plenty
On the banks of the Nacogdoches there's girls by the score
I'll wander the Brazos no more.

Refrain:
Singing la la la lee lee; give me your hand
La la la lee lee; give me your hand
La la la lee lee; give me your hand
There's many a river that courses the land

(for verse 3, substitute): The Trinity's muddy, the Brazos quicksand

Virtually all of the settlers came from places where rivers were the obvious sources of water, so some naturally chose to build their settlements near the Brazos and its many tributaries. However, due to their highly variable flow, rivers in this part of Texas often failed to provide water as a resource, but even worse, their floods could destroy and kill. Following is a summary of the history of forts, towns, and cities on the mainstem of the Brazos. Places are listed in downstream order from Possum Kingdom Lake to the Navasota River.

Brazos was named for the river and founded in 1880 near the Texas and Pacific Railway bridge. Brazos had a population of 114 people in 1940. The population then declined but rose again to almost 100 people in 2000 (Handbook of Texas Online 2009). The road bridge across the Brazos has been removed.

Dennis is an agricultural trade community located on Farm to Market Road 1543 southwest of Weatherford. It was established about 1892 after a bridge was built across the Brazos. The population has remained at about one hundred people (Minor 2010). The bridge over the Brazos remains in use.

Granbury was established in 1866 on the banks of the Brazos, although settlers, including David Crockett's widow, Elizabeth, had been in the area since the mid-1850s. Granbury became the seat of Hood County as well as an important trade center. The railroad reached Granbury in 1887, and the town continued to grow as an agricultural trade center. The downtown consists of well-preserved two-story limestone buildings, many providing attractions for the city's vibrant heritage tourism. The city's 1886 Opera House now operates as a professional live theater featuring musicals, plays, and melodramas (Minor 2010).

DeCordova Bend Dam was completed on the Brazos in 1969, resulting in the long narrow Lake Granbury that runs through the heart of the city, providing attractive lake vistas. The lake stimulated a large amount of residential and commercial development, as did the nearby Comanche Peak Nuclear Power Plant (Mayborn 2010). The population of Granbury, now 8,029, and has increased about 36 percent since 2000 (Bestplaces.net 2009).

Barnard Trading Post was established by George Barnard near Comanche Peak in Hood County in 1850 and then was moved to Waco Village in 1851. In four years Barnard shipped fifty-nine thousand pounds of undressed deer skins from his trading posts (Willingham 2010).

Glen Rose is on the Paluxy River, not the Brazos, but it is barely more than one mile from the Brazos. George Barnard, who established the Barnard Trading Post just upstream on the Brazos in 1851, opened a store on the Paluxy in 1859. Milam County officials agreed to give Barnard a section of land (640 acres) if he would build a flour- and gristmill on the river. His mill was successful, and he named the town Barnard's Mill. He sold the mill in 1871 to T. C. Jordan for $65,000, which would be $1,670,760 in today's dollars. Jordan's Scottish wife suggested changing the name. Today's Glen Rose is reminiscent of the Scottish Highlands, with its stone houses overlooking the pretty river. Barnard's Mill is now an art museum.

Glen Rose is the seat of Somervell County. In the early twentieth century people were attracted by mineral springs. During Prohibition the wooded land in the region provided cover for moonshiners, giving the area the distinction of being the "whiskey woods capital of the state" (Ferrer 2010). The Comanche Peak Nuclear Power Plant is the major employer in Somervell County. Glen Rose's population is currently estimated to be slightly less than three thousand, an increase of 37 percent since 2000 (City-data 2009).

Fort Graham was built in 1849 on the Brazos River about fourteen miles west of present-day Hillsboro. It was one in a line of forts that was to facilitate American settlement of western Texas. The U.S. Army considered Fort Graham to be one of the most important forts on the upper frontier. Because it was close to several Indian villages and an important council site it allowed the army to monitor the local Indians. Soldiers traveled north from Fort Graham to establish Fort Worth on the Trinity River. The army closed Fort Graham in 1853 as the line of defensive forts moved farther west. Its site was flooded by Lake Whitney but is commemorated by Old Fort Park on the east side of the lake (Myres 2010).

Waco is the largest city on the Brazos. It was established in 1849 at the site of a large spring on the west side of the Brazos, just south of its confluence with the Bosque River. The people that came to be known as the Waco Indians farmed the alluvial terraces above the river and maintained a village first reported by the Spanish explorers in 1541. Cherokee Indians displaced the Wacos about 1830. Texas Rangers built a short-lived fort at the village site in 1837. Land speculators with a Spanish land grant finally established the town and began selling lots in 1849 (Conger 1964, 2006).

Waco promoters induced a leading Texas figure, Shapley Ross, to relocate to the new town. One of the concessions they gave Ross was the right to operate a ferry across the Brazos, and he began the service in 1849. Ross's ferry quickly became the only good crossing on the river anywhere north of the falls near Marlin, about forty river miles downstream, and thus attracted traffic to Waco (Conger 1964). Shapley Ross's son, Lawrence Sullivan Ross, was an important leader in nineteenth-century Texas, for a time serving as president of the

Agricultural and Mechanical College of Texas, now Texas A&M University. Sul Ross State University in Alpine is named for him.

The rich blackland prairie east of Waco generated wealth through cotton production, even after the Civil War, when the slave-based cotton industry shifted to the backs of poor farmers, black and white. Immediately after the Civil War leaders in Waco began planning a bridge across the Brazos, the first and only one at the time. The 475-foot-long suspension bridge opened for traffic in 1870, firmly establishing Waco as a commercial center.

Business owners in Waco also

Even with the Brazos in flood, as shown here in 2007, Waco's suspension bridge provided the first dependable crossing of the Brazos and helped establish Waco as a trade center for the rich agricultural land of the central Brazos valley.

Chapter 5

looked to the Brazos for transportation. The riverboat captains who boasted that they could navigate a heavy dew never crawled past the Falls of the Brazos near Marlin. Since boats could not come to them, in 1874–1875 entrepreneurs built two boats at Waco with the intention of profiting by providing transportation service above the falls. The first boat, the hundred-foot *Katie Ross*, was launched in early 1875 and made several runs upstream to the Towash community, now under Lake Whitney. However, a few months later the boat's captain tried to take advantage of a rise in the river to take the *Katie Ross* to Galveston and thus earn more profit. The vessel hit hard on the falls near Marlin and was finally stranded on another shoal just downriver. Salvagers dismantled the *Katie Ross*, a sad and symbolic end for a boat named for the first Anglo child born in the Waco community,

Falls in the Brazos River, Near Marlin, Texas

The Falls of the Brazos at Marlin stymied navigation, but they have long been a popular recreation site.

This view of the Waco Suspension Bridge was mailed from Waco in 1906 to Miss Mande McIntyre in Mendon, Illinois, with no text, from "G. R. Mc." Perhaps it was a card from a traveling father to his daughter.

SUSPENSION BRIDGE OVER THE BRAZOS.   WACO, TEXAS.

the daughter of Shapley Ross. The second boat built in 1875 at Waco, the *Lizzie Fisher*, was only about half the size of the *Katie Ross*. Although promoted as a link to Towash, it disappeared in a flood. Some said wreckage was found at the falls, others said a very similar boat was later spotted in New Orleans (Puryear and Winfield 1976).

Although the *Katie Ross* and *Lizzie Fisher* failed to make Waco into a port city, city leaders continued to seek connection to the coast. A survey by the U.S. Army Corps of Engineers in 1875 documented eleven shoals below Waco that caused problems even for small rowing craft, not to mention steamboats. The survey concluded that the expense of making the Brazos navigable to Waco was unwarranted. A second survey in 1895 reached the same conclusion. Water and politics make a powerful combination, however. In 1905 the U.S. Congress authorized construction of eight dams

Head of Navigation, Brazos River, Waco, Texas.

The notation on this 1910 postcard stating that Waco was the head of navigation was incorrect because the multiple shoals downriver prevented even small boats from navigating the river with regularity. Perhaps people were looking forward to Waco becoming a port city; the U.S. Army Corps of Engineers was at that time building the dams and locks that the politicians thought would be successful, even though the engineers did not agree.

and locks below Waco to make the Brazos navigable. Construction on three of the locks and dams was under way when a major flood in 1913 destroyed the ones near Hearne and Navasota and changed the river's course, leaving the unfinished lock and dam south of Waco isolated from the river (Puryear and Winfield 1976). On Google Earth® look at the river due west of Hearne, where you will see Sullivan's Shoals, the site of Port Sullivan on the west side of the river, and an abandoned lock on the east side (30° 52' 06" N and 96° 41' 42" W). Fly to Navasota and follow Farm Road 105 west to the Brazos. Go up river three bends to 30° 23' 07" N and 96° 10' 22" W, and there you will see the dam and lock built to raise the water above Hidalgo Falls,

which is a half mile above the ruins of the lock and dam.

After its failure to become a river port, Waco literally turned its back on the Brazos, using it for waste disposal and taking notice only when it flooded. The city relied on a natural spring and drilled wells for its water supply, so the river was not important. Waco's economy developed rapidly after the Civil War, and one of the enterprising residents, Scotsman William Cameron, built a lumber empire. In 1910 Cameron's family donated 110 acres to the city as a "peoples' park" in Cameron's memory. The family later donated the grounds and clubhouse of a former country club. Ultimately the park came to total more than 400 acres, much of it bordering the Brazos and Bosque rivers.

Waco turned to the wilder but smaller Bosque River, damming it in 1929 to create Lake Waco for a municipal water supply. Although not salty like the Brazos, the Bosque carried a heavy silt load when it flooded. By the mid-1940s the storage capacity of Lake Waco had declined from about forty thousand acre-feet to about twenty-two thousand acre-feet (*Handbook of Texas Online* 2009). The U.S. Army Corps of Engineers completed a new dam in 1965. The new reservoir's storage capacity was about fifty-nine thousand acre-feet of dependable water supply. The dam's height was raised seven feet in 2003, increasing the capacity to seventy-nine thousand acre-feet (U.S. Army Corps of Engineers 2008).

Waco began to rediscover the Brazos in the late 1960s. The city had grown westward, away from the river and the old downtown. Various federal programs and local efforts ultimately resulted in a notable redevelopment of much of the riverfront within the city, including parks, the city's convention center, hotels, and

Low limestone outcroppings create shoals at several places on the Brazos as they do here at Hidalgo Falls. In the early twentieth century the U.S. Army Corps of Engineers began construction of a series of dams and locks intended to provide navigable passage over the shoals. This flood-wrecked dam was intended to raise water over the shoals at Hidalgo Falls.

Baylor University's campus includes impressive new buildings near the Brazos. The hope is that they will remain unflooded, as they did here during the record floods of 2007.

frequent floods and extended periods of low water. Instead, what appears to be the Brazos River through Waco is really Lake Brazos, impounded by a dam on the downstream side of the city. The first dam was completed in 1970 and replaced in late 2007. In addition to maintaining a relatively constant water level for aesthetic purposes, the city can withdraw up to fifty-six hundred acre-feet of water annually for municipal uses (City of Waco 2007).

Waco, especially the lower part of the city that is east of the river, has been affected by the river's floods. The east bank of the Brazos in Waco is about fifteen feet lower than the west bank, so a flood that lapped at the riverbank on the west side would do major damage on the east side. On December 10, 1913, water was twelve feet deep on Elm Street on the east side of Waco (Burnett 2008).

Port Sullivan is located on Farm Road 485 about six miles west of Hearne, just downstream from two large limestone outcrops that blocked navigation on the Brazos. It was established near Fort Sullivan, a trading post built in 1835. Port Sullivan was an active port on the Brazos from 1850 until the late 1860s, when the railroads began to supplant undependable river transportation. Its population was about fifteen hundred at its height, but nothing is left now except a marker (Brockman 2010).

Stone City is on the east side of the Brazos, just south of State Highway

the transformation of old brick warehouses into interesting restaurants. Cameron Park was given more care and attention, and a rather sophisticated zoo opened there in 1993. The zoo features its Brazos River Country exhibit, depicting features of the Brazos from the Gulf of Mexico to the

A small dam that forms Lake Brazos partially stabilizes the water level at Waco, thus providing an inviting urban waterfront. However, the Brazos does not let us forget that it will not be totally stabilized.

High Plains. The National Park Service, in cooperation with the City of Waco, Baylor University's Mayborn Museum, and the Waco Mammoth Foundation, has recently opened a facility on the Bosque River just above its confluence with the Brazos. The site contains the remains of a group of mammoths that appear to have been caught in a flood and buried under a collapsing riverbank fifty-six thousand years ago.

Waco's reconciliation with the Brazos was not done on the banks of an unfettered Brazos River, with its

21. It was founded in 1892 at the terminus of the Hearne and Brazos Valley Railway and was located at the site of Moseley's Ferry, which served the area in the 1860s. In 1914 it had several stores, a post office, and about thirty residents. By 1940 there were about fifty residents. Today only a few dwellings remain (Odintz 2010).

THE DAMMED BRAZOS RIVER

As noted above, the first dams on the Brazos were begun in 1905 below Waco, as part of an unsuccessful U.S. Army Corps of Engineers project to make the river navigable to Waco. However, in response to destructive floods on the lower river, regional, state, and federal organizations began to discuss the desire for dams in the middle part of the river, where the narrow valleys offered efficient storage space to retain floodwaters (Committee on Flood Control 1919). In 1902 local leaders formed the Brazos River Improvement Association, which developed into the Brazos River and Valley Improvement Association in 1915 and finally into the Brazos River Conservation and Reclamation District in 1929, renamed the Brazos River Authority in 1953. It was the first river basin management agency in the country and became a model for the Lower Colorado River Authority and the Tennessee Valley Authority. The agency's water development and management responsibilities were mandated in order of priority: domestic and municipal, industrial, irrigation, mining, hydroelectric power, navigation, and recreation (Hendrickson 1981). It is notable that flood "control" was not a priority.

Consistent with the country's attitudes about water in the first half of the twentieth century, the Brazos River managers were dedicated to building dams to "conserve" water and to prevent it from being "wasted" by flowing into the Gulf. A history of the Brazos River Authority commissioned and published by the authority in 1981 states that "in spite of wide fluctuations from one to fifteen million, an average of six million acre-feet of water flowed through the Brazos each year, but most of it was wasted into the Gulf of Mexico. Particularly at its peak stage the river was a water waster" (Hendrickson 1981, 62).

In 1935 the Brazos River Conservation and Reclamation District adopted a master plan developed by Ambursen Engineering Corporation, a New York company that had built hundreds of large dams worldwide. The plan called for six dams on the mainstem: Possum Kingdom, Turkey Creek, Inspiration Point, DeCordova Bend, Bee Mountain, and Whitney, plus seven dams on tributaries (Hendrickson 1981). In 1954 the authority proposed its Six Dam Plan, which would have produced a 250-mile chain of reservoirs, from South Bend, above Possum Kingdom, to Whitney, including the three additional dams between Possum Kingdom and Whitney at Turkey Creek, Inspiration Point, and Bee Mountain (Hendrickson 1981). This was the plan that motivated John Graves to make his farewell trip on the Brazos and write *Goodbye to a River*.

The Texas engineering firm Lockwood & Andrews reviewed the Six

MAJOR RESERVOIRS ON THE BRAZOS MAINSTEM

Dam (construction date)	Capacity (acre-feet)	Surface area (acres)	Length (miles)	Maximum width (miles)	Maximum depth (feet)
Possum Kingdom (1941)	Conservation pool: 540,340 Flood pool: none	17,624	27	1.8	145
Granbury (1969)	Conservation pool: 128,046 Flood pool: none	8,310	25	0.9	75
Whitney (1951)	Conservation pool: 553,344 Flood pool: 1,300,000	23,220	20	3.6	108

Dam Plan and recommended the project:

> The initial phase of the proposed development, including water conservation and power development by the construction of five new dams between Possum Kingdom and Whitney Dams, raising Possum Kingdom Dam, and with certain changes in operation of the Whitney project, is considered to be a worthy project, based on the assumption that economic feasibility will be established by others. There appears to be ample future market for the water supply and power produced. This portion of the river offers the most economical dam sites, sufficient water resources for further development, and certainly the most logical locale for the development of hydroelectric power. (Lockwood & Andrews 1955, 149)

However, the engineers made it clear that water from the project would have a limited number of uses due to the high natural salt content of the upper Brazos:

> Under present conditions of quality of water, there appears to be no assurance that the quality of water, from the initial phase of 6-Dam Development, will be satisfactory for municipal and industrial use, although indications are that water from the 6-Dam Development will be better than present Possum Kingdom water in quality. Also, in view of the present lack of quantitative data on both natural and artificial pollution, there appears to be no assurance that it will be found feasible to reduce this pollution sufficiently to insure that a firm supply of good quality water satisfactory for municipal and industrial use can be supplied from the initial development, without further dilution downstream. (Lockwood & Andrews 1955, 150)

The Brazos River agency, under its various names, struggled to realize its ambitious plan, seeking funding from federal and state agencies and at times using its own fund-raising authority. The Brazos River Authority did not build all of the dams it considered, but with the cooperation of the Corps of Engineers it generally carried out the spirit of the 1935 master plan. It took sixty years to do so, but that is amazingly fast for water projects.

There are 1,178 sizable dams in the entire Brazos basin and almost innumerable stock tanks, which are usually earthen dams on small drainages that also retain water from the river. Twenty dams with at least 5,000 acre-feet of storage capacity were built in the Brazos basin before 1960 and sixteen since then. The current storage capacity in the Brazos basin is 4,613,800 acre-feet (Dunn and Raines 2001). Nine potential new dams are listed in the 2007 Texas Water Plan.

Possum Kingdom, Granbury, Whitney, and Lake Brazos are on the mainstem of the river. Lake Brazos at Waco is intended to hold water in the channel, maintaining a very slow-moving river at a relatively constant water level. The three larger reservoirs upstream more clearly represent the effects of major dams on the Brazos.

The U.S. Army Corps of Engineers built four reservoirs in the Brazos system upstream of Waco, all with large flood pools to detain floodwater. Lake Whitney is the only flood control reservoir on the mainstem and has an immense flood pool of 1.3 million acre-feet. Since Whitney Dam was completed in 1951 the Brazos at Waco has flowed at 100,000 cfs only once, when the great drought broke in 1957 (U.S. Geological Survey 2008). Aquilla, Proctor, Waco, Belton, Stillhouse Hollow, Georgetown, and Granger reservoirs are also part of the system of reservoirs planned since the 1930s to contain floods on the Brazos. For example, Lake Waco has a flood pool of 553,300 acre-feet compared to its conservation pool of 79,000 acre-feet. The conservation pool is Waco's primary water supply, and the large flood pool detains floods from the incredibly "flashy" North Bosque.

The flood detention dams on the Brazos passed a major test in 2007. Record rainfall began in March 2007, and by July the Brazos watershed had received its total average annual rainfall. Because Possum Kingdom Lake and Lake Granbury have no flood retention capacity the man-

agers had to release water quickly. Possum Kingdom released 450,000 acre-feet of water from mid-June to early July, almost the total capacity of the reservoir. Lake Granbury released an astonishing 805,000 acre-feet of water, more than six times its capacity. This water flowed into Lake Whitney and was detained in the reservoir's 1.3-million–acre-foot flood pool. Lake and system managers were confident but carefully working to balance releases from Whitney with those downriver at Lake Waco. The astounding thing is that the peak flow of the Brazos at Waco was only 39,900 cfs during this time. Lake Whitney and the other three flood detention reservoirs detained 2.5 million acre-feet of water during the spring and summer of 2007. By the fall of 2007 the reservoirs were back to their normal levels. The system worked just as the engineers planned it in the 1930s (Brazos River Authority 2007).

Corps of Engineers reservoirs on tributaries of the flood-prone Little River provide examples of the complex issues involved in building reservoirs and the evolution of agency policy and water management priorities. Those reservoirs include Belton, completed on the Leon River in 1954; Stillhouse Hollow, on the Lampa-

Access is somewhat difficult along much of the flowing Brazos, but the lakes provide easy access that provides recreation and beauty for many people.

sas River in 1968; Georgetown, on the North Fork of the San Gabriel in 1978; and Granger, on the San Gabriel in 1977. In *Road, River, and Ol' Boy Politics* Linda Scarbrough presents a fascinating and detailed analysis of the conflicts between and within water management agencies and local political factions regarding Georgetown and Granger dams. But she also traces the evolution of people's expectations of major water projects. All four of the dams in the Little River system were authorized and constructed for what was once called "flood control." Regarding reservoirs on Brazos tributaries,

Such reservoirs could create transportable liquid pools for desperate situations in the high-priority downstream world of rice and Dow. As a BRA spokesman put it, "You could go to another bucket that was full—or fuller." This was the sub rosa reason the BRA made a case for two, or even three, dams on the tiny San Gabriel, and why the Texas Water Commission and politicians like Lyndon Johnson and Jake Pickle backed them up. Sympathy for Williamson County and its flooding miseries [was] the cover story that helped sell the project, but by the sixties, greater forces—Dow and the rice industry—were Texas policy-makers' larger concerns. (Scarbrough 2005, 166)

However, there was little consensus about where the Georgetown and

Granger dams should be located. The political wrangling that began in the 1930s continued until 1966, when local congressional representative J. J. "Jake" Pickle deftly worked through a compromise.

PEOPLE AND THE LAKES

The stretch of the Brazos that John Graves immortalized in *Goodbye to a River* was already regulated by Possum Kingdom Dam upstream and Whitney Dam downstream. Graves wrote clearly about the presence of those dams, but somehow we seem to forget that and have taken his book to be the farewell to a natural river. The idea that *Goodbye to a River* protests against dams but recounts a trip on a dammed river expresses our ambiguity about dams. Some people recognize that dams can truly damn rivers, but most of us love the lakes they create and think not about the rivers they impede. Woody Guthrie, that most populist of populist singers, in 1941 wrote songs extolling the virtues of dams. Of Bonneville Dam on the Columbia River he said that its power is "bringing darkness to dawn," and "shiploads of plenty will steam past the docks" (Murlin 1991, 14–15).

Of course, his dam songs were commissioned by the Bonneville Power Administration and were essentially propaganda, but they did express sentiments that were strongly felt at the time and that millions of Americans still feel.

The lakes changed the river, but

they have been there for decades now and have formed their own ecosystems, including various exotic species that we value highly, such as the striped bass. The benefits that originally justified the dams perhaps were illusory. On the mainstem reservoirs the water is too salty for municipal use or irrigation. The electricity generated is not a significant part of the state's power supply. Most of the dams in the Brazos system were built by the Corps of Engineers to detain floods, which they do. However, since 1972 the national policy has been to prevent flood damage by not building in the floodplain rather than by building flood detention dams. East Waco is the main beneficiary of flood detention by several of the dams, and that part of the city was developed long before the policy makers changed their minds. For us today it is a moot point because the dams are in place.

Recreation was the lowest priority for the Brazos River Authority, although by the 1950s the river planners did recognize its importance. As it has worked out, recreation may be the most important benefit of the dams and lakes. Texas does not have much land that is accessible to the public, and the large tracts that are available, such as Big Bend National Park, are far removed from the population centers. But consider just the Corps of Engineers lakes on the Brazos. Those lakes provide more than seventy-five thousand acres of water surface, hundreds of miles of shoreline, plus scores of parks and hundreds of campsites, all accessible to the public and within short distances of most of the residents of Texas. John Graves (2002, 9) wrote disparagingly of "Chriss-Crafts [*sic*] and the tinkle of portable radios" on the new lakes. On the other hand, the lakes have provided millions of hours of recreational time for families and individuals.

One might question if it is possible to have a "nature" experience on an "unnatural" reservoir, and for some people the answer would be negative. Reservoir levels fluctuate more than natural lake levels do, so the shoreline vegetation does not develop well. Reservoirs are more turbid than natural lakes. Many people are sensitive to what was lost when the river was covered by the impounded water. On the other hand, like a natural lake, a reservoir is a topographic depression that holds water, thus providing opportunities for a variety of life. In addition to offering opportunities for fishing, Texas reservoirs attract many birds and are popular for bird watching. Of course, a whole range of outdoor activities can be enjoyed at the many parks on the Brazos lakes. The main advantage of the lakes in terms of recreation is that they make the water easily accessible. The experience may not be as profound as being on a natural river, but it can be valuable nevertheless.

Motorboating is popular on Texas lakes, but only about 20 percent of Texans participate in motorboating and 10 percent use personal watercraft (Cordell 2004). A study in 2006 of boaters on Possum Kingdom Lake found that the average boat on the lake ranged between nineteen and twenty-four feet long, with motors from 200 to 270 horsepower (Titre et al. 2006). Such boats cost thirty thousand to eighty thousand dollars new but much less if previously owned. The costs of fuel, insurance, and maintenance are also substantial, which may explain why such a small percentage of the population participates in motorboating. An infinitesimal 4 percent of Texans sail, but there are some interesting trends that may change the way we use lakes in Texas, especially the lakes on the Brazos.

Kayak fishing is a recent phenomenon in Texas that is growing rapidly. Many anglers have realized that they can buy a new fishing kayak for fifteen hundred dollars. These boats are easier to transport and store than motorboats. Kayaks allow their users to fish in areas that would be difficult to access with a motorboat. Larry Bozka (2007) has written about using kayaks in the Texas bays, but anglers who frequent the lakes are recognizing the same advantage that he discovered:

It was a real epiphany when about a year ago, having just turned 50, I realized that my evolution as a boater and fisherman had quite possibly come full-circle. Like most other

anglers, my fishing career began on the bank and soon graduated to a small aluminum john boat. From then on, I craved, and ultimately ran, bigger and bigger hulls with progressively larger engines and more sophisticated accessories with bigger and bigger price tags.

Now, at the phase of life when so many things that once obsessed me suddenly seem senseless, even silly, simplification has become my prime directive. Way beyond needing to impress myself or anyone else with high-dollar toys, I now spend as much time as possible fishing from a boat that's smaller and lighter than anything I have owned.

Kayaks are inexpensive and safe when used properly. They offer increased opportunities for water recreation on the Brazos and its lakes, such as this sheltered cove at Possum Kingdom Lake.

It's also a lot less expensive. Fifteen hundred bucks doesn't go far on an outboard rig, but it'll buy a top-notch, well-appointed fishing kayak.

There is another important potential for "re-creation" on the Brazos lakes, as well as other lakes in Texas. One of the characters in the British children's book *The Wind in the Willows* said, "There is nothing . . . half so much worth doing as simply messing about in boats" (Grahame 2003, 13). Published in England in 1908, this book was not referring to 200-horse-power speedboats but to simple row-boats and sailboats. Today in several parts of the United States there is a movement called "Family Boating," which involves building and using small rowing and sailing boats. The *Community Boatbuilding Manual* says that small boats help us learn self-reliance, teamwork, forethought, honesty, endurance, tolerance, and modesty (Youcha 1998). Small boats can help people learn safety and responsibility, boat handling, navigation, and maintenance. Boating on a lake can teach about weather, wildlife, geology, and water science. The Brazos lakes are within a short drive of millions of Texas families who might enrich their lives by "messing about" in a small boat.

Homes with a lake view are another highly valuable result of the lakes. A sample of twenty homes for sale at Possum Kingdom Lake listed on the Internet in July 2009 had a me-dian value of $1.6 million per home, with the lowest price home offered at $890,000 and the highest offered at $3.5 million. By contrast in Graham, Texas, the nearest community to Possum Kingdom Lake, the median home value in July 2009 was $52,840. Clearly, the lake views and access are quite valuable. Boathouses and docks are allowed on Possum Kingdom and Lake Granbury. While such structures are nice for the owners and generate high economic value, they clutter the shoreline and diminish its beauty.

FLOATING THE RIVER BETWEEN THE LAKES

Between Possum Kingdom Dam and Waco are about 230 miles of un-dammed river, excluding the lengths of lakes Granbury and Whitney. John Graves wrote about part of this, from Possum Kingdom to the upper end of Lake Whitney before Lake Granbury was built.

The Brazos River between the lakes is one of the most popular places in Texas for paddling and fishing. The water is relatively clear because the sediment is deposited above the dams. The fine sediments in the river are washed away, so there are gravel banks and bars, plus pools and riffles. In places there are lime-stone cliffs adjacent to the river. The Brazos here fulfills our stereotype of what a river should look like. The Nationwide River Survey described the reach between Possum Kingdom and Lake Granbury as follows:

Within migration route of Whooping Crane, a federally listed endangered species. Recommended for inclusion in proposed Texas Natural Rivers System. Rated as #1 scenic and recreational river in the northern half of state by River Recreationist Survey; one of top 10 in the state. Flow regulated by Possum Kingdom Dam, often only 20 cfs, but is heavily used for recreation. Barber Mountain–Pollard Bend area is one of the state's 100 top rated natural areas because of scenic, vegetation and wildlife values. Texas Natural Areas Survey indicated that rare plants occur at springs along the bluffs. Boy Scouts use area extensively for training and badge qualifications. (National Park Service 2009, n.p.)

The water is usually low, and the current is slow. There are no public parks on these reaches of the river, but there are private lodges, fishing camps, and RV parks that provide access. Some of them rent canoes and kayaks and provide shuttle service. Search the Internet for "Brazos River cabins" to find the currently available sites. Fishing, especially for bass and catfish, is popular on these reaches. There are several professional guides on these segments of the Brazos. Search the Internet for "Brazos River Texas Outfitters."

John Graves Scenic Riverway

John Graves was Texas' first notable environmental activist, as his book protested the Six Dam Plan. It is ironic that the Brazos River should be the birthplace of Texas environmentalism given that even John Graves said few people would really care about the river. Fortunately, there are people who do care about the Brazos, especially the part that Graves traveled. Although the reach of the river between Possum Kingdom and Whitney lakes is not natural, it is certainly a pleasant river and has attracted many visitors and residents.

Because the Brazos is a large and flood-prone river, over its long history (until the dams) it transported huge amounts of sand and gravel, much of which was deposited in its shifting floodplain. The sand and gravel are now an important resource for the rapidly growing Fort Worth and Dallas area. Riverbanks and floodplains have long been the primary sources of sand and gravel. However, the mining operations release large amounts of sediment into the river. There has been little regulation of the environmental impacts of sand and gravel mining in Texas, so there are few laws that concerned citizens can invoke to remedy the problem. However, some of the people attracted to this part of the river have possessed substantial political influence and prompted the Texas Legislature to designate the John Graves

The sections of the Brazos River below its three mainstem dams are popular paddling routes. In Goodbye to a River *John Graves protested additional dams that would have made the river into a chain of reservoirs from Possum Kingdom downstream to Whitney*

Scenic Riverway. This legislation imposed regulations on sand and gravel operations in Parker and Palo Pinto counties. The Brazos River Conservation Coalition emerged as a result of the process and continues to monitor conditions. The Friends of the Brazos is another environmental organization that works to restore a healthy river environment by encouraging the various agencies that regulate the river to maintain essential instream flow.

Hydropower generation ceased at Possum Kingdom in 2007, thus reducing releases into the river. Given that fact, as of late 2010 the Brazos River Authority is studying new operational protocols for Possum Kingdom Lake, Lake Granbury, and Lake Whitney. While there are many demands for the water that was previously used for hydropower generation, I and many others would hope that the water will be used to help maintain the proper volume and timing of sustainable instream flow in the river.

Where to Experience the Land and the River

There is a profusion of places to experience the land and water of this section of the Brazos. There are ten Corps of Engineers lakes that offer recreation areas and nine state parks plus two wildlife management areas. In Waco there are views of the Brazos and Bosque from Cameron Park, paddling trails along both rivers, the impressive Brazos River Country exhibit at the Cameron Park Zoo, and the Lake Waco Wetlands. The following state parks offer particularly good access to the land and water of the central Brazos watershed.

POSSUM KINGDOM STATE PARK: P.O. BOX 70, CADDO, TEXAS 76429
940-549-1803

Possum Kingdom State Park is in the Cross Timbers and Prairies ecoregion. The park offers campsites and cabins overlooking the clear water of Possum Kingdom Lake. The lake is located where the Brazos cuts through the Palo Pinto Mountains, so there are beautiful cliffs at the edge of the lake and deep coves that are delightful for kayak touring. The park provides easy access to Possum Kingdom Lake, which is popular for fishing, boating, and scuba diving.

The park helps us understand the rugged topography of this part of the Brazos basin and illustrates how the river was able to cut through limestone and sandstone. The oldest rocks in the Brazos basin are Pennsylvanian period limestones, located toward the eastern end of the lake but not inside the park. The park also makes abundantly clear the high salt content of the Brazos with many posted signs reminding visitors that the water is not potable.

LAKE MINERAL WELLS STATE PARK AND TRAILWAY: 100 PARK ROAD 71, MINERAL WELLS, TEXAS 76067
940-328-1171

Lake Mineral Wells State Park and Trailway are in the Cross Timbers and Prairies ecoregion. The lake was completed in 1920 to supply water for the city of Mineral Wells. In 1925 the Texas National Guard established Camp Wolters near the lake. Until 1974 it was the site of a military training facility, one that offered helicopter training. The land and lake were given to the Texas Parks and Wildlife Department in the mid-1970s.

Lake Mineral Wells State Park demonstrates the importance of surface water in the Brazos basin. The city of Mineral Wells built the lake for water supply based on the Conservation Amendment of 1917. Although no longer the city's main water supply, the lake continues to be an important recreational resource. In 1998 the Texas Parks and Wildlife Department opened the Lake Mineral Wells Trailway, which follows the old Texas and Pacific railroad route for twenty miles between Weatherford and Mineral Wells. The trail is unusual in that it provides a way to have a more intimate experience with the land of the Brazos basin than is usually available.

DINOSAUR VALLEY STATE PARK: P.O. BOX 396, GLEN ROSE, TEXAS 76043
254-897-4588

Dinosaur Valley State Park is on the Paluxy River, a tributary of the Brazos that drains the Cross Timbers and Prairies ecoregion. The park gives us a glimpse of a life drama from 113 million years ago. Exposed in the river's bed was a double set of tracks in the ancient Glen Rose limestone. One set consisted of saucerlike impressions more than three feet long and two feet wide, with a stride of seven to ten feet. The second set of tracks appeared to follow the first. Those tracks were birdlike but as large as twenty-four by seventeen inches, with a stride of up to sixty-five inches. Paleontologists surmised that the larger tracks were from a herbivore thirty to fifty feet long called a *Pleurocoelus* and that the smaller tracks were from a carnivorous dinosaur, perhaps an *Acrocanthosaurus*. The tracks suggested that the carnivore was chasing the larger dinosaur. Paleontologists calculated their speeds at 5 and 2.7 miles per hour, respectively, so it is possible that the smaller animal, the carnivore, dined well that day.

Dinosaur tracks are visible in the limestone bed of the Paluxy River at Dinosaur Valley State Park

As was common in the early twentieth century, paleontologists from the American Museum of Natural History in New York excavated and removed the rocks with the sequence of tracks, giving some of them to the Texas Memorial Museum in Austin. Some of those are now on display in the interpretive center at Dinosaur Valley State Park. Although the dramatic scene is no longer in the river, you can see similar tracks in the riverbed plus the more rare tracks of what was probably a thirty-foot-long *Iguanodon*.

Dinosaur Valley State Park has been designated by the National Park Service as a National Natural Landmark and is a wonderful place to learn about the river and the land. The little Paluxy River is usually clear and delightful for wading on the limestone and gravel bottom and for looking for the whole range of river creatures. Because the river flows through relatively dry countryside, the vegetation zonation from riparian to bottomland to upland is obvious. A trail along the river has a set of signs that explain the fluvial processes of this flashy river. There are two fiberglass model dinosaurs—a seventy-foot one that is thought to be similar to the *Paluxysaurus* and a forty-five-foot *Tyrannosaurus rex*, a larger relative of the *Acrocanthosaurus*.

CLEBURNE STATE PARK: 5800 PARK ROAD 21, CLEBURNE, TEXAS 76033
817-645-4215
Cleburne State Park is located in the Cross Timbers and Prairies ecoregion, within a valley of springs that has long been recognized as a special and pleasant place. Comanches and cattle drivers camped here. In

the 1930s the Civilian Conservation Corps built a variety of structures, including a dam that formed a small lake that is now the central feature of the park.

Cedar Lake is unusual in the Brazos valley because it is fed by springs, and consequently it is cool and clear. Because the lake is only about 116 acres, motorboats are required to go very slowly. Thus, the lake is especially good for canoeing and kayaking. Fishing and bird watching are popular activities.

MERIDIAN STATE PARK: 173
PARK ROAD 7, MERIDIAN,
TEXAS 76665
254-435-2536
Meridian State Park centers on seventy-two-acre Lake Meridian, which was formed by a dam built on Bee Creek by the Civilian Conservation Corps in the 1930s. The CCC also built a limestone concessions building overlooking the lake and a trail that encircles the lake. The landscape is dominated by white limestone covered by juniper and oak and provides habitat for the endangered golden-cheeked warbler. The park is very representative of the Cross Timbers and Prairies ecoregion.

The lake is good for fishing and canoeing or kayaking. The trail provides views of the lake and the surrounding countryside from tall cliffs. This small park offers quiet, intimate relationships with the landscape of the Brazos watershed.

LAKE WHITNEY STATE PARK:
BOX 1175, WHITNEY, TEXAS
76692
254-694-3793
Lake Whitney State Park, along with the Corps of Engineers campgrounds, provides ample access to Lake Whitney and is representative of the Cross Timbers and Prairies

ecoregion. Perhaps unique in a state park, this park provides a two-thousand-foot runway for aero-campers. Because Lake Whitney was built primarily for flood detention the Corps does not allow private homes and boathouses at the shoreline, which is one of the factors that makes the lake so pretty compared to many other lakes in Texas that are lined with houses, boathouses, and docks.

Whitney is a popular fishing lake, considered excellent for smallmouth bass, white bass, striped bass, and catfish. Notable records from Whitney are a sixty-three-pound blue catfish that was forty-six inches long, a forty-pound striped bass forty-three inches long, and a sixty-five pound grass carp forty-eight inches long. Sailing is popular on Lake Whitney because it catches the prevailing south wind. The large number of sheltered and secluded coves makes Lake Whitney suitable for kayak touring.

MOTHER NEFF STATE PARK:
1680 STATE HIGHWAY 236,
MOODY, TEXAS 76557-3317
254-853-2389
Mother Neff State Park was the first state park in Texas, built on a six-acre site on the Leon River donated in 1916 by Isabella Eleanor Neff, mother of Texas governor Pat M. Neff. The Civilian Conservation Corps built facilities, including a limestone water tower with an observation platform on top. The site was originally named Mother Neff Memorial Park

and was officially opened as a state park in 1937 after the CCC facilities were completed.

This park is located in the Cross Timbers and Prairies ecoregion and is useful for understanding the flood power of rivers in the Brazos system. Normally the Leon River looks inconsequential, but when it is in flood it spreads widely across the landscape, including much of Mother Neff State Park.

FORT PARKER STATE PARK:
194 PARK ROAD 28, MEXIA,
TEXAS 76667
254-562-5751
Fort Parker State Park is especially useful for setting the historical context of the Brazos and emphasizing the large size of the Brazos basin. The park is in the Blackland Prairie ecoregion. It was built to commemorate the Indian raid in 1836 when Comanches captured Cynthia Ann Parker and took her to the region of the upper Brazos tributaries two hundred miles or more to the west, where she later became the mother of the last Comanche war chief, Quanah Parker. The original wood stockade fort was destroyed, and a replica was built in the 1930s. The replica is not in Fort Parker State Park but is near-

The Texas Parks and Wildlife Department provides popular and high-quality access to lakes on the mainstream of the Brazos and its tributaries.

by in a park operated by the City of Groesbeck.

The Civilian Conservation Corps built a dam on the Navasota River that formed a seven-hundred-acre lake in the park. The Navasota is the easternmost tributary of the Brazos, so the landscape of Fort Parker State Park is representative of true prairie lands.

LAKE SOMERVILLE STATE PARK AND TRAILWAY AND WILDLIFE MANAGEMENT AREA:
BIRCH CREEK, 14222 PARK ROAD 57, SOMERVILLE, TEXAS 77879-9713
979-535-7763

NAILS CREEK & TRAILWAY, 6280 FM 180, LEDBETTER, TEXAS 78946-7036
979-289-2392
Lake Somerville State Park and Trailway is a large park with two sections connected by a thirteen-mile-long trail. The Birch Creek section is on the north shore of the lake, and the Nails Creek section is on the south shore. The trailway loops around the west side of the lake through the Lake Somerville Wildlife Management Area. Side trails provide an additional seven miles of wilderness hiking. The park spans both the Blackland Prairie and Post Oak Savannah ecoregions.

Lake Somerville is popular for the entire range of water sports. However, the outstanding feature of this area is the large amount of land accessible to the public. The two sections of the park comprise 5,520 acres, and the wildlife management area is 3,180 acres, for a total of 8,700 acres of wild land. Much of this area is bottomland, which attracts a wide variety of species. The area includes Flag Pond, which is a 350-acre natural depression that provides rich wetland habitat.

CHAPTER 6

The (Almost) Free Brazos

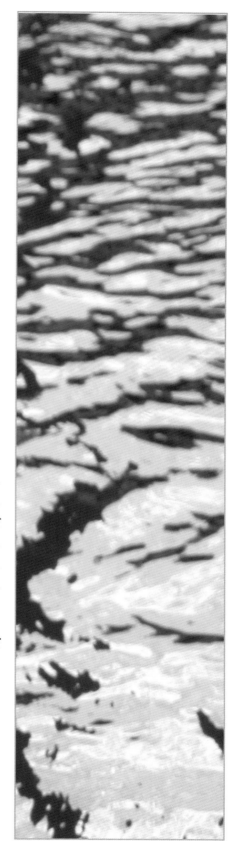

The bottom of the Brazos in the neighborhood of San Felipe is about seven miles wide, which compares in width with the Mississippi at St. Louis. The richest and most fertile soil which Texas possesses, suitable for raising cotton, sugar cane and corn, lies in the Brazos bottom, particularly below San Felipe.

Ferdinand Roemer, *Texas* (1849)

THE PREVIOUS CHAPTER describes people's efforts to make the Brazos River behave as they wanted. This chapter focuses on a very different river, one that is surprisingly natural and free—one of the longest stretches of (almost) free river in North America. In *America by Rivers* Tim Palmer listed what he termed "long, undammed sections of rivers." The lower Brazos is the seventeenth longest undammed section of the 155 river sections Palmer identified.

This part of the Brazos is one of the most historic regions of Anglo American Texas. It is also a region of subtle beauty, as illustrated in *Reflections of the Brazos Valley* by Gentry Steele and Jimmie Killingsworth (2007). Thus, the message of this chapter is that the lower Brazos is indeed special to Texas as both a remnant of our natural heritage and a remembrance of our cultural heritage.

The map shows this section of the Brazos to begin below the mouth of the Navasota. However, the (almost) free river actually begins below the major dams where the rivers begin to redevelop their natural processes. Thus, we have a mainstem of the almost free river beginning below Lake Brazos Dam at Waco, about four hundred river miles from the Gulf. This

The (Almost) Free Brazos

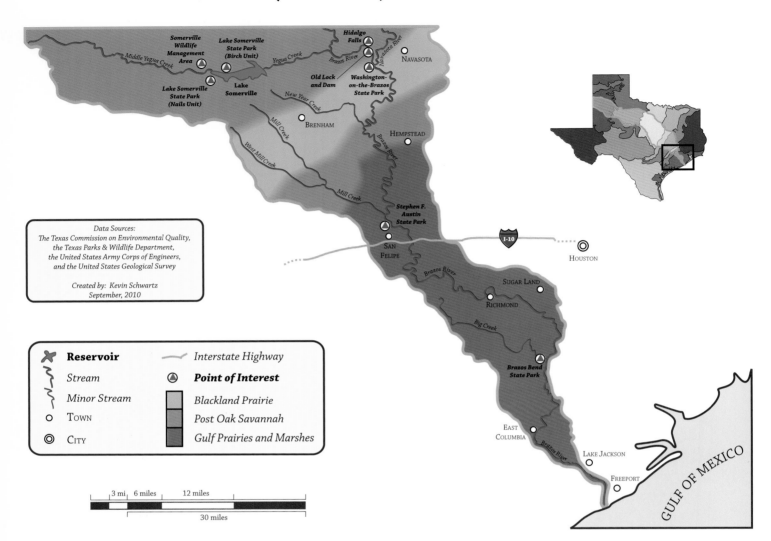

Data Sources:
The Texas Commission on Environmental Quality,
the Texas Parks & Wildlife Department,
the United States Army Corps of Engineers,
and the United States Geological Survey

Created by: Kevin Schwartz
September, 2010

is 50 percent longer than the Hudson River and twice as long as the river Thames. There are also substantial tributaries that run free below their dams, such as almost one hundred miles of the Navasota and more than ninety miles of the Little River.

Is it legitimate to think of this part of the Brazos as an almost free, almost natural river since there are so many dams above it? Certainly it is not truly a natural river. The dams trap sediment, reduce flood heights, and reduce flow due to evaporation and other losses. They interrupt natural processes and timing described by the river continuum concept. On the other hand, four hundred miles of river is sufficient to re-create natural processes. As we will see, even the sediment load of the river is not significantly less than the pre-dam load.

Water of the Almost Free Brazos

CLIMATE

This part of this Brazos catchment basin receives a substantial amount of rainfall. The average annual rainfall at College Station is forty inches. Just east of the mouth of the Brazos, at Galveston, it is forty-four inches. Temperatures are moderate to warm, ranging from an average low at Galveston of 66 degrees Fahrenheit to an average high at College Station of 79 degrees Fahrenheit. However, there are extremes of rainfall and temperature. Both College Station and Galveston have had one-day rainfalls of thirteen to fourteen inches. The record high temperature in the region was 112 degrees Fahrenheit at College Station (National Oceanic and Atmospheric Administration 2010a and 2010b). What this means for the Brazos is that the deep soils, warm temperatures, and abundant rainfall support lush vegetation that helps absorb rainfall and moderate runoff to the river.

FLOW

The average annual flow near the mouth of the Brazos at Richmond, Texas, is 7,585 cfs and the maximum was 123,000 cfs on June 6, 1929, although an unconfirmed flow of 300,000 cfs was reported in 1913. Dr. John Washington Lockhart described more historic variations in the Brazos's flow:

My father moved to Texas and settled on the banks of the Brazos river in 1840, just above the town of Washington. Just previous to this time there had been one of the greatest overflows of the river that had ever been known. The water extended from the town of Washington to Ringold's prairie, near the present city limits of Navasota. The ferryman, in his crossing, had marked the trees in order to keep his

course. The marks were visible for years afterward, and showed that the water was from ten to fifteen and in some low places twenty feet high in the bottoms.

After this rise for a succession of years the country was very drouthy. It was very hard to raise a crop. We had to plant corn in January in order to catch the very limited spring rains. . . . In consequence of this drouthy condition the river got to be very low. It looked for a time that it would stop running.

In 1840 or 1841 Mr. Tillotson Wood, now living at Hempstead, and myself rode from Washington to Hidalgo falls, some seven miles by land, and perhaps double that distance by the river, along the bed of the river, never once having to leave it, but crossing from the point of one sand bar to the other. In some places it looked as if a ten foot rail would reach across the water. (Wallis 1967, 82)

Lockhart and his family missed the 1833 flood and then watched as the Brazos seemed about to go dry. Then in 1842, as it did in 2007 after a long dry spell, the Brazos produced one of its largest floods:

During this great rise of the river my father gathered his corn crop in boats, the water having covered the stalks of corn as far up as the ears. This was in June, and it was late in July before the flood subsided. During this great overflow many buffaloes, wild horses

and smaller game were seen floating down the river intermixed with driftwood which consisted of immense cottonwood, oak, ash and pecan trees, green and fresh, going to show that there had been great erosions of the banks of recent date. In fact when the water got near the top of the banks the soil, being so loose and loamy, sometimes as much as one-fourth of an acre would slip into the river. (Wallis 1967, 83)

The upstream dams have moderated fluctuations in flow, but in spite of the upstream dams, a recent study by the Texas Water Development Board states that "the flow regime in the lower Brazos River basin has remained similar to the historic flow regime primarily because the nearest on-channel reservoir, Lake Whitney, is located several hundred kilometers upstream" (Osting, Mathews, and Austin 2004, 6). What has changed since the dams were completed is that higher floods are less frequent because several of the dams are intended to detain flood water (Osting, Mathews, and Austin 2004).

WATER QUALITY

While the Brazos from Waco to its mouth is not dammed and thus has many characteristics of a natural river, it is not pristine. There are many, varied, and widely dispersed sources of pollutants. This part of the Brazos watershed includes all or parts of Washington, Grimes, Austin, Waller,

The Brazos from Waco to its mouth is almost natural—four hundred river miles of wildness through the heart of Texas.

Fort Bend, and Brazoria counties. The first four are rural counties with populations ranging from about twenty-six thousand to about thirty-seven thousand. Fort Bend County has a population of more than half a million and Brazoria County has about three hundred thousand people, but only about 15 percent of the county is in the Brazos watershed. In total, there are almost seven hundred thousand people in this part of the watershed, plus part of one of the largest petrochemical installations in the world. Many of the homes are in rural areas and use on-site wastewater treatment.

Near its mouth at Richmond, the Brazos carries more than twenty-three million tons of sediment each year. That amount has not significantly changed since the dams were

The humid subtropical climate of this region of Texas supports rich vegetation and provides a radically different environment for the Brazos compared to that of its upper reaches.

Rivers continually erode and deposit bank materials as floods rise and recede.

built, because the water that flows through the dams contains only a small amount of sediment and is able to erode the banks and bed (Dunn and Raines 2001). The 2007 Texas Clean Rivers Program report noted that the Brazos from the mouth of the Navasota to the Gulf of Mexico has bacterial and nutrient pollution from urban, industrial, and agricultural runoff and effluent. Oyster Creek, a major tributary near the mouth, has elevated levels of nutrients, including nitrate-nitrite and phosphorus, and bacteria, including *E. coli* from fertilizer, municipal discharges, and malfunctioning on-site sewage facilities.

Land of the Almost Free Brazos

GEOLOGY AND TOPOGRAPHY

The entire lower Brazos is within the Gulf Coastal Plain physiographic province, which ranges in age from Early Tertiary in the north to Pleistocene along the coast (about 60 million to 100,000 years old). These materials were brought in and deposited by the Brazos and other rivers and consist of fine sands, silts, and clays that form an almost level landscape. However, beneath the surface are Jurassic period salt domes from 144 million to 208 million years ago that have been pushed to the surface, plus Oligocene

coral formations that are 24 million to 37 million years old (Jordan 1984; Spearing 1991).

SOILS, VEGETATION, AND LAND USE

The river in this section crosses three ecoregions: the Blackland Prairie, Post Oak Savannah, and Gulf Coast Prairies and Marshes. Although the Post Oak Savannah was originally a diverse area of woodland and grassland, it and the two prairie ecoregions have been almost completely converted to improved pasture and intensive cultivated agriculture.

After Anglo settlement in the early to mid-1800s much of the land was converted to row crop cultivation; however, from 1924 to 1992 the production of non-hay crops in the region decreased by 32 percent, so that now only about 8 percent of the land is cultivated in row crops, including cotton, corn, sorghum, and rice near the coast. The rest is in improved pasture and hay, both of which subject the land to less erosion (Dunn and Raines 2001). On Google Earth® go to College Station, Texas, the home of the state's primary agricultural research university, and fly around, noting the extensive amount of agriculture along the river. Petroleum production is important in this region as well.

Ecology of the Almost Free Brazos

RIVER CONTINUUM CONCEPT

As outlined in chapter 2, the river continuum concept hypothesizes that natural rivers change from narrow shaded headwaters, where energy and materials mostly come from outside the system, to the central part of the river that is wider, possibly clearer, and produces most of its energy and materials via photosynthesis, to the lower part of the river, which is likely to be too turbid for local photosynthesis to be greater than the energy and materials that are washed into the river.

Does this almost free river that begins below the dams share characteristics with an undammed river? The water below the dams is relatively clear because its sediment was deposited above the dams, and the water is well oxygenated due to the turbulent flow from the dams, so in these respects it may be similar to the headwaters of a natural river. However, the river below the dams is not narrow, as is the case with the hypothesized upper river in the river continuum concept, and thus is not shaded, nor does it receive a substantial amount of energy and materials from riparian vegetation, as would be the case with the upper part of a natural river. Thus, it is incorrect to think that the upper reaches of this undammed part of the Brazos function in a way similar to the upper reaches of the hypothetical

Although the Brazos is not as clear as many would like, Hidalgo Falls above the mouth of the Navasota River is a popular recreation site.

The Blackland Prairie, Post Oak Savannah, and Gulf Coast Prairies and Marshes provide distinctly different landscapes for the (Almost) Free Brazos and add to its appeal as a natural area of great importance to Texas.

"natural" river in the river continuum concept.

The Brazos quickly picks up sediment below Waco, so this part of the Brazos is turbid and light does not penetrate very deeply into the water, so photosynthesis is limited. The lower Brazos receives runoff from its narrow valley as well as the long undammed portion of the Navasota. Thus, this part of the Brazos probably does function as the lower part of a river is hypothesized to function by the river continuum concept.

FLOOD PULSE CONCEPT

Although the upstream dams have reduced the frequency and height of floods on the lower Brazos, the flood pulse concept seems to continue to be relevant here because the river maintains hydrologic and biological connections with its floodplain and, most importantly, with the oxbow lakes in the floodplain. Of six oxbow lakes studied recently the river connects with three of them more than once a year, one at least every two years, one about every four and a half years, and the last only during high-magnitude floods (Osting, Furnans, and Mathews 2004; Zeug, Winemiller, and Tarim 2005).

LOCAL RIPARIAN PRODUCTION

As you fly over the lower Brazos in Google Earth® you will note that almost all of the banks are lined with trees, even where the land has been converted to pasture or row crops. Thus, there should be substantial amounts of materials that fall into the river from this riverbank vegetation.

How "natural" is the Brazos from Waco to the Gulf? Am I justified in calling it an almost free river? Recent studies by fisheries biologists at Texas State University–San Marcos offer some (almost) encouraging insights. Comparing characteristics of the river before and after most of the dam construction, the scientists found that flood frequency and volume had declined, but floods still occur. They found that the average flow actually has increased slightly in the lowest part of the river (Bonner and Runyan 2007).

Regarding the ecology of the river, the recent studies indicated that eight species declined from 1939 until 2006 and four increased. However, of the other fifty-five species, the population trends of thirty-nine were stable and those of sixteen were indeterminate. Non-native species were only a minor component of the lower Brazos fish community (Bonner and Runyan 2007). Whether or not the

lower Brazos is almost natural depends on perspective. The authors of the study state that "we consider [the] lower Brazos River fish assemblage imperiled because of the number of endemic and semi-endemic forms that are decreasing in abundance. Yet, if we exclude consideration of these taxa (or they never existed), the lower Brazos River fish assemblage would appear exceptional because the majority of the fish assemblage is intact" (Bonner and Runyan 2007, 20). Yes, the Brazos has changed ecologically since 1939, probably as a result of human action, but it is still a viable river ecosystem worthy of our continued care and protection.

Other than insects, the most common animal in the (Almost) Free Brazos is a minnow called the red shiner (*Cyprinella lutrensis*). In fact, more than 90 percent of the fish in the lower Brazos are of the *Cyprinidae* family (Thomas, Bonner, and Whiteside 2007, 33; Texas Parks and Wildlife Department 2009). However, the presence of the alligator gar fish and the American alligator reptile emphasizes the fact that this part of the Brazos is still a wild place.

ALLIGATOR GAR (*ATRACTOSTEUS SPATULA*)

The record alligator gar caught in the Brazos thus far is 7.75 feet long, weighing 190 pounds. However, alligator gar may be longer than my 9-foot kayak and may weigh almost 300 pounds (Texas Parks and Wild-

Nature certainly captures our attention when it is awesome and maybe even a little frightening. The oxbow lakes at Brazos Bend State Park provide opportunities to observe alligators closely, but safely.

life Department 2010a; Biology Department Texas State University–San Marcos 2010). Alligator gar have a double row of teeth in their long jaws and prey on other fish, crabs, and birds. They also eat refuse. Alligator gar live in slowly moving water, pools, backwaters of the river, and oxbow lakes. Other than one minor incident years ago on the Rio Grande, there are no records in Texas of alligator gars harming people (Texas Parks and Wildlife Department 2010b).

Gar have traditionally been considered a "trash" fish, not good to eat, but that prejudice is changing. Alligator gar are now classified as a game fish by the State of Texas. Anglers catch them on rod and reel and with bow and arrow. Surely, catching such a fish is incredibly exciting—exciting enough that a fishing guide service advertises three days of alligator gar fishing on the Brazos for $2,450. Although the guide service does not offer a guarantee, it indicates that 100 percent of its clients catch at least one gar of one hundred pounds or larger (FishQuest 2010).

AMERICAN ALLIGATOR (*ALLIGATOR MISSISSIPPIENSIS*)

Based on "nuisance" reports the Texas Parks and Wildlife Department

has documented that alligators live as far up the Brazos as Waco, but their greatest concentration is from East Yegua Creek and the Navasota downstream to the Gulf. Their name comes from the Spanish *el lagarto*—the lizard. The record alligator in Texas is fourteen feet, four inches, and nine hundred pounds, but the Texas Parks and Wildlife Department says that most alligators one might see in Texas are only five to six feet and weigh twenty-five to fifty pounds. Alligators are not particular about what they eat: "Alligators greater than 4' long eat a wide variety of food items and are very opportunistic. The diet of an adult alligator is mostly made up of crawfish, crabs, non-game fish, and carrion. Occasionally, game fish, large turtles, wading birds, waterfowl, muskrat, nutria, otters, raccoons, alligators, feral hogs, and white-tailed deer are eaten. Non-food items, such as shotgun shells, glass bottles, brass objects, and wood are frequently consumed by alligators, so it is important not to litter" (Texas Parks and Wildlife Department n.d.).

Alligators are lethargic and avoid people. But if they are threatened, especially a female at her nest, they can move quickly, faster than we can run or paddle our kayaks. Females build their nests in May and June, making the mounded nest out of vegetation, bark, cattails, and mud. Clutches of eggs average thirty-five in number, and they hatch about sixty-five days after they are laid. This means that the mother will begin protecting her young in late August through early September. The Texas Parks and Wildlife Department says that there has never been a recorded death of a human due to an alligator attack in Texas, but there have been several in Florida. However, it was reported that the servant of the French explorer La Salle was pulled from their raft and killed by an alligator when La Salle's group was crossing the Brazos in 1686 (Campbell 2003).

Seeing an alligator in the wild is one of those humbling yet exciting experiences that remind us that we are still quite tender, regardless of our technology and power. That realization is one of the reasons that the almost free lower Brazos is valuable to us. Brazos Bend State Park, described later, is a good and safe place to see alligators.

People and the (Almost) Free Brazos

PREHISTORY

As we might expect given the widespread distribution of the Clovis and other Paleoindian peoples, Clovis-style projectile points found along stream courses indicate that people were living on what is now the Texas coast ten thousand to eleven thousand years ago (Ricklis 2004). However, it is impossible for archeologists to thoroughly research where these people might have lived because the sea level rose about three hundred feet from the end of the last ice age (about ten thousand years ago) until about three thousand years ago, when it stabilized (Ricklis 2004). As in the rest of the Brazos valley, the Clovis people were hunters and gatherers but were particularly skilled in killing large animals. The type of stone used for the projectile points found on the lower Brazos is not from the region, indicating that these early people moved into the area from the Great Plains, where such stones are found (Ricklis 2004), or perhaps traded for the stones.

Not as old as the Clovis artifacts but certainly thought provoking are two cemeteries located on the lower Brazos. Actually, the cemeteries are in one location but separated by depth. The oldest may be as much as 4,200 years old and the later one as recent as 1,500 years old. The older site contains remains of 61 people, most with their heads oriented to the southeast. The newer cemetery contains the remains of 145 people, most oriented to the northeast (Ricklis 2004). Is there a possibility of some cultural connection, given that the site was twice used for burial purposes, even though almost 3,000 years separated the two occasions?

The people living along the shoreline and the rivers as far back as 7,500 years BP may have had a relatively pleasant life. One archeologist has stated that they had a "superabundance" of food resources, including nuts, white-tailed deer, and a great variety of estuary and river species (Ricklis 2004). However, during that time there were two periods of about 1,000 years each when the populations appear to have been much smaller (6,800 to 5,900 BP and 4,200 to 3,100 BP). Archeologists think these may have been periods of more rapid sea level rise, which reduced the productivity of the estuaries due to increased salinity (Ricklis 2004). Of course, there is also a superabundance of mosquitoes in those marshy places.

HISTORIC INDIANS

The Brazos may have been an "ethnic boundary" between the famous Karankawa and the similar but less well-known Akokisas (Ricklis 1996). The Karankawa lived down the coast from the Brazos, while the Akokisas and other groups lived northeastward into what is now Louisiana (La Vere 2004). When Cabeza de Vaca landed on what was probably Galveston Island in 1528 he noted that there were two Indian groups there, the Cavoques and the Han. The Cavoques were one of the Karankawa groups and the Han were part of the same group as the Akokisas (La Vere 2004).

Stephen F. Austin, who seemed always to be presented as a rather mild man, said of the Karankawa, "These Indians may be called universal enemies to man . . . their [sic] will be no way of subduing them but extermi-

nation" (quoted in Campbell 2003, 115). His prescription was ultimately carried out, as the Karankawa had been destroyed as a people by 1858 (La Vere 2004).

However, these people that lived for perhaps two thousand years on the richness of the bays and rivers of the coast probably were not the abject savages they are reputed to have been. Although they were reported to practice cannibalism, Cabeza de Vaca recounted that the people he encountered were outraged when some of the starving Spaniards ate each other (Cabeza de Vaca 2002). Cabeza de Vaca wrote that after he was shipwrecked for the second time, in November 1528, "Upon seeing the disaster we had suffered, our misery and misfortune, the Indians sat down with us and began to weep out of compassion for our misfortune. For more than half an hour they wept so loudly and so sincerely that it could be heard far way" (Cabeza de Vaca 2002, 33). Cabeza de Vaca lived with these people for six years, sometimes as a slave, sometimes as a trader and medicine man, but they did not eat him.

EUROPEANS

Cabeza de Vaca was the first European to report about the Brazos region, one of the areas through which he wandered during an odyssey that began in 1528. He was lost, so his geographic descriptions are subject to speculation, but his chronicle pro-

vides fascinating insight into the native peoples of the area. Many years later, in the late 1600s, as the Spaniards were establishing missions in East Texas, their route crossed the Brazos near the mouth of the Little River or at the site of the later Port Sullivan. The Spanish explorers apparently had no other relationship with the Brazos from that point downstream. They established no missions in the region. About the time it gained independence from Spain, the Mexican government built a fort named Quintana on the west side of the mouth of the Brazos, just as the Americans began to arrive.

Growing up in Texas I learned about the Texan heroes at school and from my grandmother, whose ancestors were part of both Austin's and DeWitt's colonies. Austin, Crockett, Fannin, Lamar, Travis, and even my own Davis ancestors unquestionably did heroic deeds. What did not quite dawn on me, however, was why they were here in the first place. The romantic story is that they were seeking human liberty in the style of the Jeffersonian yeoman farmer. The more rational interpretation is that they were primarily attracted by economic opportunity. The Austin family of Virginia and then Missouri had been wealthy until the economic collapse of 1819. Moses Austin recognized the potential to rebuild the family fortune by becoming an empresario in Texas, and he put things in motion to do so. He died in 1821 and his son,

Stephen Fuller Austin, took over the enterprise, which was focused on facilitating the development of cotton and sugar plantations (Campbell 2003). For the time and place it was a rational business plan. The Mexican government authorized the Austins to recruit settlers who would pay the Austins for their land. Most of the settlers, including my ancestors, were Jeffersonian-style farmers, but subsistence farming does not create wealth. The real economic engine of Austin's enterprise on the Brazos was the one-third of the landowners who developed slave-based plantations (Campbell 2003). The plantations produced valuable molasses and cotton, both of which were labor intensive and relied on large numbers of slaves to clear the land and plant, harvest, and prepare the crops for market. Slaves often outnumbered their white owners (Kelly 2010).

NAVIGATION ON THE BRAZOS

The Austin land grant embraced both the Colorado and Brazos rivers. Having lived and owned industries in Connecticut, Virginia, and Missouri, Moses Austin must certainly have understood the importance of rivers to the economic development he planned for his colony in Texas. For each family Austin planned to allot 320 acres of river frontage and 640 acres off the river (Campbell 2003). Austin was influenced to select the Brazos as his main focus by the colorful former tax collector Philip

Mary Austin Holley's Brazos Boat Song

Stephen F. Austin's cousin, Mary Austin Holley, visited Texas via the Brazos River in 1831 and composed the following song.

> *Come whistle my boys, to the good San Antonio**
> *Whistle my boys, that fav'ring gales blow*
> *Come whistle my boys, to the good San Antonio*
> *Whistle my boys, that fav'ring gales blow*

> *Bright shines the morning Sun, Long ere the day is done,*
> *We'll moor in our forest home, Far o'er the wave*
> *We'll moor in our forest home, Far far o'er the wave*

> *Come whistle my boys, to the good San Antonio*
> *Whistle my boys, that fav'ring gales blow*
> *Spread wide the white sheet, Make fast the tall sprit*
> *And steer for our forest home, Far o'er the wave.*

> *Come whistle my boys, to the good San Antonio*
> *Whistle my boys, that fav'ring gales blow*
> *And as we chant our song, Swiftly we glide along*
> *To our forest home vespers, Far, far o'er the wave.*

*Saint Anthony was the patron saint of riverboat men (Mooney 1998).

Mary Austin Holley's lyrics do not mention the Brazos; they pray for winds suitable for sailing craft, which were not used on the Brazos, and they refer to a forest home, but the Texas coast was primarily prairie. Was she using poetic license, or was she simply very homesick?

Hendrik Nering Bögel. Bögel had re-created himself in Texas as the "Baron de Bastrop" because he was wanted in his home country of the Netherlands for embezzlement (Campbell 2003). The baron advised Austin well, because the Colorado River was blocked by a massive logjam and was not navigable. However, as we will see, the Brazos was not to become the Rhine River of Texas.

Thus began decades of efforts to navigate the Brazos, finally ending with the pork-barrel locks and dams discussed in the previous chapter. Although historians who analyzed Brazos navigation have concluded that "the Brazos was simply not designed by nature to accommodate steamboats," the struggles to navigate the Brazos demonstrated the tenacity of people's commitment to their life in Texas (Puryear and Winfield 1976, xiii).

For a very short time in the river's long life it was the primary artery of transportation for a large portion of the settled parts of Texas and in fact was the major entry point for early American settlers coming into the state. More than seventy steamboats navigated the Brazos between 1830 and 1895 (Puryear and Winfield 1976). The boats were both sternwheelers and sidewheelers. The sternwheelers were best for the river because the sternwheel, put into reverse, would force water under the hull if the boat ran aground. The sidewheelers were best for the open water run to Galveston. Some of these boats were more than 120 feet long and could carry 300 bales of cotton.

Navigating the Brazos was a problem even at its mouth. We have already seen that the river's heavy silt load filled its bay. That load continued to settle at the river's mouth, forming a shallow clay barrier called the Velasco Bar, which blocked all but shallow-draft boats at high tide. Later, this bar inspired the construction of what is now the Gulf Intracoastal Waterway.

The Brazos was part of a high-stakes power game between Galveston and Houston in the mid-nineteenth century, as upstart Houston tried to displace wealthy Galveston as the commercial center of Texas. Houston bet on the combination of its ship channel through Galveston Bay and its connection with the yet-to-come railroads. Galveston tried to connect itself to the interior of the state via the Brazos River, reached by a canal through West Galveston Bay and the low wetlands between the bay and the Brazos. The Brazos Canal Company started digging a canal in 1847 using slave labor, but the project failed after taking two years to dig one mile. The Galveston and Brazos Navigation Company completed a canal in 1855 from San Luis Bay to the Brazos River, following creeks and bayous where possible and cutting across land where necessary. The canal was designed to be three feet deep, but that depth was not always maintained. The canal avoided the Velasco Bar and allowed the river sternwheelers to go to Galveston without having to go out into the open water of the Gulf, where waves could wreck their sternwheels. The idea of this canal, if not its specific location, ultimately became the Intracoastal Waterway but not until long after the 1900 hurricane had devastated Galveston and Houston's rail connections had established its primacy (Puryear and Winfield 1976).

The only commercial shipping use of the Brazos now is in its old mouth. Fly to Freeport on Google Earth® where you can see the twisty old mouth going southeast and the new river channel going almost due south (28° 57' 44" N, 95° 22' 34" W). Note the petrochemical plants served by the Dow Canal. Zoom in and look at the long berm on the river's east side that diverts the flow into the new channel.

The old river mouth is now the entrance to the Port of Freeport, the fourteenth largest port in the United States in foreign tonnage (Port of Freeport 2009). The channel into the port is dredged to maintain a four-hundred-foot width and forty-five-foot depth. The Velasco Bar that caused so much trouble is gone.

TOWNS AND CITIES ON THE BRAZOS

Except for the port cities of Galveston and Houston, virtually all of the towns and cities of early Anglo American Texas were located on the Brazos because it was the main transportation artery. The following descriptions are arranged from upstream to downstream, not chronologically.

Washington, now known as Washington-on-the-Brazos, is located on the right bank of the Brazos across the river and just downstream from the mouth of the Navasota near the place where an Indian trail that the Spanish explorers called La Bahia Road crossed the Brazos. By 1822 a ferry had been established there and

The original mouth of the Brazos was cut off in the 1930s when the river was diverted and is now the entrance to Freeport Harbor. Frequent maintenance dredging is necessary to maintain a navigable channel.

the town was laid out in 1833, protected from floods by its location on the high west bank.

The town became a trade center, and during the Texas Revolution it was Sam Houston's headquarters. In 1836 it played host to the convention to establish the new Republic of Texas. Sawmills and a brickyard supplied materials for the town's rapid growth, and by 1839 the population was about 250 people. Washington was the capital of the Republic of Texas from 1842 through 1845. It was the capital when the Texas congress approved annexation to the United States.

The river steamer *Mustang*

worked its way to Washington in 1842. John Lockhart lived in Washington then and described the arrival of the *Mustang*: "Somewhere early in the 40s, I do not now remember the year, a small steamboat called the 'Mustang' came up to Washington on, I presume, an experimental trip, for I do not remember her having much freight aboard. I think it was a trip merely to ascertain the feasibility of the navigation of the river. I remember distinctly the stir it created among the citizen[s] along the valley, and especially the citizens of the town and country about Washington" (Wallis 1967, 84). The success of the *Mustang* motivated community leaders to develop the town as a river port, even to operate steamboats themselves. Lockhart described the venture:

> The trip of the "Mustang" so encouraged the citizens of Washington that they employed Mr. Y. M. H. Butler of Galveston to go to Pittsburg, Penn., to have two steamers built of light draught to run as regular packets on the Brazos.
>
> They were christened "Brazos" and "Washington," but made only two trips each. By some means they became so heavily encumbered with debt that they had to be sold, and were put into the trade of the lower Brazos. (Wallis 1967, 86)

In spite of the problems that Washington business owners had with their own boats, river commerce was the mainstay of economic activity for them. Perhaps because they had such a commitment to river transport, in 1858 they refused to pay eleven thousand dollars to connect Washington to the Houston and Texas Central Railroad, and the railroad went to Hempstead instead. Washington began to decline rapidly as residents moved to Brenham or Navasota. Even the Episcopal church building was moved to Navasota (Christian 2009). The site of the once important town is now the location of Washington-on-the-Brazos State Historic Site.

San Felipe de Austin was founded in 1824 as the administrative center of Austin's Colony, the first major Anglo settlement in Texas. Austin chose the site where the Old San Antonio Road crossed the Brazos (Moore 2009).

There was a ferry at the river. The town site was on high bluffs, protected from floods. There were springs and a creek for water supply. By 1828 the town had become a trade center as well as the administrative center for the colony. About two hundred people lived in San Felipe in 1828, approximately 90 percent of them men. Cotton plantations and stock grazing were the basis of the local economy. Keelboats provided transportation to and from the mouth of the Brazos eighty miles downstream. However, the flow was undependable and the route circuitous, so overland transportation was preferred. After 1830 steamboats were common on the Brazos (Jackson 2009).

By 1835 San Felipe's population was about six hundred and the town was second only to San Antonio as a trade center for Texas. Given its importance as an administrative center for the Anglo settlements in Texas, San Felipe was an important meeting place for those fomenting the revolution against Mexico.

San Felipe residents fled after the fall of the Alamo in 1836 and General Houston's retreat. The town was burned by the fleeing residents. After the Texan army defeated Santa Anna some residents returned to San Felipe, but the destroyed town never regained its economic and administrative importance. The German traveler Ferdinand Roemer described San Felipe ten years later:

> The city of San Felipe de Austin, indicated in large letters on all maps of Texas, is made up of from five to six miserable, dilapidated log and frame buildings. One of these is a combination store and saloon. In front of this building were a half dozen men, who drank whiskey freely, and whose haggard frames and pale faces gave plain evidence of over-indulgence in whiskey and the ravages of fever. Otherwise there were no activities or signs of life visible and we were glad when this dismal, deserted place was left behind us. At one time, San Felipe was supposed to have had six hundred inhabitants, but, during the

last war between Texas and Mexico when the Texan army retreated, the commander in charge of the Texans ordered it burned to prevent it from forming a base for the Mexicans. Since that time the place has not been rebuilt, owing, probably, to its unhealthful location near the Brazos bottom where huge quantities of vegetable matter are constantly decaying. (Roemer 1935, 77)

San Felipe revived somewhat but never regained its original status. Like Washington, it was bypassed by the railroad, which established a station at Sealy, and people moved there from San Felipe. In the late 1800s a railroad spur was built one-half mile south of the old town of San Felipe and residents moved to that location (Jackson 2009). Today San Felipe has a population of about one thousand people (Best Places 2009a). Much of the old town site is now within Stephen F. Austin State Park.

By betting their futures on river transportation rather than the railroads, Washington and San Felipe truly were going against the flow. Not only was rail transport more dependable but the Texas Legislature offered huge subsidies to railroad companies

Beginning in 1822 a ferry crossed the Brazos from the east bank (midground in photo) to Washington, now Washington-on-the-Brazos State Historic Site. The steep, often muddy banks surely were an inconvenience.

beginning in 1854. After the railroad company laid a minimum of twenty-five miles of track it received sixteen sections of land (10,240 acres) for each of those miles and for each additional mile of track laid. As if this were not enough, in 1856 the legislature offered loans of six thousand dollars per mile (Campbell 2003). **Richmond**, unlike Washington and San Felipe, survived as a community because it ultimately cooperated with its nearby rival, Rosenberg. Anglo settlers first formed a community on a big bend on the Brazos in about 1822 and built a fort called Fort Bend. Like the other Texan communities, it was evacuated during the Runaway Scrape in 1836, but it was re-established as Richmond in 1837 and became the seat of Fort Bend County.

Rail connection was established from Houston to Richmond in 1855, and the town prospered as a trade center. However, in 1878 Richmond refused right-of-way to a railroad that ultimately became part of the large Santa Fe Railway, so the railroad company founded Rosenberg three miles west of Richmond (Leffler 2009; Myers 2009; Werner 2009).

By 1884 Richmond's population was up to two thousand, but it declined to half that by 1904 as Rosenberg and other communities attracted residents. A wooden bridge was built over the Brazos about 1888. It collapsed a few years later and a steel bridge was built. Crossing the Brazos at Richmond on the modern U.S. 90

bridge gives a sense of how deeply incised the river channel is and how powerful the constrained flow would be during a flood.

Richmond's current population is about thirteen thousand and Rosenberg's is about thirty-three thousand (Best Places 2009b, 2009c). **Fulshear** is at the site of a plantation established by Churchill Fulshear as part of Austin's Colony. The Fulshear family operated a cotton gin and flour mill and gave right-of-way to the San Antonio and Aransas Pass Railway in 1888. People moved to Fulshear from nearby Pittsville when the railroad was built. The community functioned as a small agricultural trade center with a population of a few hundred (Crawford 2009).

Although the original community was not on the Brazos, modern subdivisions have been built near oxbow lakes formed by the Brazos and its tributaries and some of the homes are near the Brazos.

Sugar Land is east of Richmond and is fifteen to twenty feet lower than Richmond in elevation, which makes a major difference in its relationship to the Brazos. The river is not so deeply entrenched and consequently can more easily flood the surrounding land. Sugar Land originated as a twelve-thousand-acre sugar plantation and ultimately developed into a company town for the Imperial Sugar Company, owned by the Kempner family of Galveston. In the early 1900s the company built more than eight miles of levees and twenty miles of drainage ditches to prevent flooding from the Brazos. By 1928 the company had stopped growing its own sugarcane and converted the land to grazing. Through the 1930s and 1940s the population was about fifteen hundred. In 1959 the Kempner family began to sell land and businesses, and the city of Sugar Land was officially incorporated. The rapid growth of Houston created demand for housing, and the former plantation lands have been intensively developed behind levees to protect them from Brazos floods. Sugar Land currently has more than eighty thousand residents (Anhaiser 2009; Best Places 2009b; Kleiner 2009c).

Sugar Land is developing a park along nine miles of the Brazos River, including about thirty-six hundred acres of floodway. Sugar Land Memorial Park, the first phase of the project, opened in 2007 (City of Sugar Land 2007).

East Columbia and West Columbia were founded by plantation owner Josiah Bell. East Columbia was founded in 1823 as Marion and served as the headquarters for Bell's plantation. It provided access to the Brazos and was also called Bell's Landing. The population reached its peak of fifteen hundred in 1890 and declined afterward to less than one hundred. The town was named East Columbia in 1927.

West Columbia, founded simply as Columbia in 1826, was more successful than Bell's venture at Marion. It was located just south of Varner's Plantation, which became one of the largest sugar plantations in the state and is now the Varner-Hogg Plantation State Historic Site. Although not located on the Brazos, Varner's Plantation was located on Varner's Creek, which enters the Brazos at East Columbia. During the Texas Revolution in 1836 the town served for a short time as the capital of the Republic of Texas. It hosted the republic's first congress, President Sam Houston was inaugurated there, and Stephen F. Austin died there in December 1836.

West Columbia fluctuated in business activity and population, but in 1918 oil was discovered nearby and the town became economically viable. West Columbia now has an estimated population of about four thousand (City-data.com 2009a; Jones 2009b; Weir and Kleiner 2009; Weir 2009a).

Brazoria was founded in 1828. Located about twenty-five river miles from the Brazos's historic mouth at Velasco, it is the first river town upstream. It functioned as a river port and had about eight hundred residents by 1884. As with many of the river communities, the railroad bypassed Brazoria and its development faltered. However, the Farm Road 521 bridge over the Brazos at Brazoria provided connection to the rapidly developing Houston metropolitan area to the east, and Brazoria's population has grown to about three thousand (Hallstein 2009; City-data.com 2009b).

Lake Jackson began in 1943 as a company town for the new Dow Chemical plants that were being built around the old mouth of the Brazos at Freeport. The city was named for the oxbow lake that had been part of the sugar and cotton plantation built on the site by Abner Jackson in 1843. Lake Jackson's population grew rapidly along with the expansion of the Dow manufacturing facilities, increasing from 2,897 residents in 1950 to 11,090 in 1958. The city also developed as a regional trade and professional center. The estimated population in 2006 was 27,614 (Rice 2009; U.S. Bureau of the Census 2009a).

Although not a historic river town like Brazoria or Richmond, Lake Jackson provides important opportunities to gain understanding and appreciation of the region. The Jackson Plantation Archeological Site is a Texas Antiquities Archeological Landmark located on the shores of Lake Jackson (Few 2007). Sea Center Texas is a facility of the Texas Parks and Wildlife Department, described in more detail later. Lake Jackson's Wilderness Park is almost five hundred acres, with frontage on both Buffalo Camp Bayou and the Brazos River. The park has hiking trails and abundant wildlife, including alligators.

Freeport was established in 1912 by the Freeport Sulphur Company to

Levees built to protect sugarcane fields from Brazos floods now guard homes for thousands of people at Sugar Land.

The City of Freeport provides access to the old river channel at City Park, which includes a public boat ramp. **Quintana** was the location of a Mexican fort built soon after Mexico gained independence from Spain in 1821. It is on the right bank of the old river mouth, now separated from the mainland by the Gulf Intracoastal Waterway. The waterway was completed to Corpus Christi by 1941.

Quintana was an important port in early Texas. The trading firm of McKinney, Williams and Company was established at Quintana in 1834 and became a dominant commercial force in Texas, focusing on banking, real estate, and steamboat transportation on the Brazos and other Texas rivers. Quintana was a popular resort and vacation center before the Civil War. A Confederate fort there was bombarded by the U.S. Navy. Galveston supplanted Quintana's port functions, as well as Velasco's on the other side of the river. The hurricane of 1900 destroyed everything on the coast. Quintana is now the site of beach houses, a large shipping terminal, and a county park (Bishop 2009; Jones 2009a).

Old Velasco was located at the mouth of the Brazos on the left bank of the river opposite Quintana. Some sources state the old town was four miles from the Gulf, but historic maps clearly show the town on the Gulf. The schooner *Lively* landed the first of Austin's colonists at the site in 1821. It became the major port of

provide shipping access for what became one of the world's largest sulfur mines. However, the harbor was the narrow, twisty mouth of the Brazos, which often flooded. Floods and the sediment they deposited caused problems for shipping. In the 1930s the federal government funded a project to redirect the river and relocate its mouth about six miles farther down the coast, directing floodwaters away from what is now the busy Port of Freeport (Townsend 2007). In 1939 Dow Chemical began development

of its magnesium and other chemical production facilities in and near Freeport, using water from the Brazos for part of its production processes (Kleiner 2005; Scarbrough 2005). Dow Chemical and the Brazosport Water Authority have permits to withdraw about 350,000 acre-feet per year from the Brazos for industrial, municipal, and irrigation use, including providing water to Angleton, Brazoria, Clute, Freeport, Lake Jackson, Oyster Creek, and Richwood and the prison units in Brazoria County.

Quintana and Velasco at the original
mouth of the Brazos served as ports
throughout the nineteenth century.
However, Freeport Harbor was able to
be developed more intensively after the
mouth of the river was relocated to carry
floodwaters directly to the Gulf.

The somewhat difficult access to much
of the lower Brazos adds to its sense of
wildness, as seen here, close to Lake
Jackson.

Fishing Boats, Freeport, Tex.

The old mouth of the Brazos has long been a port for commercial and sport fishing boats.

entry for Austin's colony and the transshipment point connecting the Brazos River with the ports of Galveston and New Orleans. In 1831 the Mexican government established a customs house and fort there. A dispute over customs tariffs resulted in the Battle of Velasco in 1835 between Anglo Texans and the Mexican soldiers stationed there, one of the events leading to the Texas Revolution.

After the Texas Revolution Velasco served as the temporary capital of the Republic of Texas and General Santa Anna signed the Treaty of Velasco there, recognizing the new republic. Between the revolution and the Civil War Velasco took on the character of many port cities, including relative affluence and an almost cosmopolitan ambience. There were many businesses, and no doubt there were many ruffians and other crude characters. However, there were summer resorts and houses for wealthy planters and a racetrack, plus

the Velasco Female Academy and a boy's school with faculty who had been educated at Oxford University (Weir 2009b).

When the Galveston and Brazos Navigation Company opened its canal to Galveston in 1856 both Velasco and Quintana began to fade as shipping ports, but their functions as places of leisure for the wealthy plantation families continued. However, the Civil War and the end of the plantation economy brought that activity to a halt. During the Civil War Velasco defended the mouth of the Brazos River to prevent federal ships from taking on fresh water and supplies. A hurricane in 1875 did major damage, and by the late 1880s the town barely functioned (Weir 2009b).

However, waterfront property motivated promoters in the 1890s as well as today, and a new Velasco

Sea Battle at the Mouth of the Brazos

On April 17, 1837, a sea battle took place just off the coast from Velasco. As described in an official report, a Texan ship was captured and taken to a naval base at Brazos Santiago, near Matamoros, Mexico, where the report was written:

Brazos de St Iago April 21st 1837 To the Honorable S Rhoads Fisher, Secretary of the Navy

Sir—I have the honor hereby to transmit you an account of the late engagement between our government vessel Independence *and two of the enemy's brigs of war, one the* Libertador *of sixteen eighteen pounders, 140 men; the other, the* Vincedor del Alamo, *mounting six twelve-pounders, and a long eighteen amidships, with one hundred men. Captain Wheelright having during the action received a very dangerous wound, the duty of sending this melancholy communication has devolved upon me, to wit:*

On the morning of the 17th, in latitude 29 deg. N., longitude 95 deg. 20 min. W., at 5 h. 30 in A. M. discovered two sail about 6 miles to windward; immediately beat to quarters; upon making us out they bore down for us with all sail set, signalized, and then spoke each other. At 9 h. 30 m., the Vincedor del Alamo *bore away, getting in our wake to rake us, the* Libertador *keeping well on our weather quarter, we immediately hoisted our colors at the peak. The enemy in a few minutes hoisting theirs, the* Libertador *on our weather quarter edging down for us all the time, till within about one mile, gave us a broadside, without wounding any of our men or doing other damage; the fire was at the same time returned from our weather battery, consisting of three sixes and the pivot, a long nine, the wind blowing fresh, and from our extreme lowness our lee guns were continually under water, and even the weather ones occasionally dipped their muzzles quite under. The firing on both sides was thus briskly kept up for nearly two hours, the raking shots from the* Vincedor *in our wake nearly all passing over our heads, as yet sustaining but trifling injury; at 9 h. 30 m. the* Libertador *on our weather quarter, bore away and run down till within two cables length of us, luffed and gave us a broadside of round shot, grape and canister, while all this time the brig* Vincedor *in our wake continued her raking fire. Notwithstanding this we still continued on our course for Velasco, maintaining a hot action for full 15 minutes, with some effect upon her sails and rigging. The* Libertador *now hauled her wind, widening her distance, apparently wishing to be further from us, when she again opened her fire, which was on our part kept up without cessation. At 11 A. M. she again bore away, run down close to our quarter and gave us another broadside of round shot, grape and canister, which told plainly on our sails and rigging; as before she again hauled her wind to her former position, and played us briskly with round shot, one of which struck our hull, going through our copper and buried itself in her side. At 11 h. 30 m.*

A. M. a round shot passed through our quarter gallery, against which Captain Wheelright was leaning, inflicted a severe wound on his right side, knocked the speaking trumpet out of his hand, terribly lacerating three of his fingers; he was conveyed below to the surgeon, leaving orders with me to continue the action. We still held on our course in our respective positions, keeping up an incessant fire, for full half hour, when the enemy signalized; then the Vincedor *in our wake luffed up and gained well on our weather quarter; at that time the* Libertador, *on our weather beam bore away and ran down under our stern within pistol shot, our decks being completely exposed to her whole broadside, and at the same time open to the raking fire of the* Vincedor *on our weather quarter. In this situation, further resistance being utterly fruitless, and our attempts to beach the vessel ineffectual, I received orders, from Captain Wheelright to surrender, which was done.*

The only damage done to our vessel, was that of parting some of our rigging, splitting the sails, a round shot in her hull, and the quarter gallery, which was shot away. Captain Wheelright was the only person wounded on board. We shot away the Libertador's *main top-gallant mast, unshipped one of her gun carriages, took a chip off the after part of the foremast, killed two men, and cut her sails and rigging severely. We were immediately boarded by capt Davis of the* Libertador, *who pledged his honor, and that of Commodore Lopez, who was then on board, that we should receive honorable treatment as prisoners of war, as officers and gentle-*

men, and as soon as an exchange could be effected, we should be sent home. The kind attention and courtesy we have received from Commodore Lopez, Captain Davis and officers has been truly great for which we tender them our sincere thanks, likewise Captain Thompson of the schooner of war Bravo has extended every civility and kindness. We leave this place tomorrow for Matamoras: what disposition will be made of us I know not.

Besides the officers and crew of our vessel, we had on board as passengers, the honorable Wm. H. Wharton, Mr. Levy, Surgeon T. N., captain Darocher, T. A., Mr. Thayer, of Boston, Mr. Wooster, English subject, George Mess, acting lieutenant T. N. and Mr. Henry Childs.

I remain very respectfully, your obedient servent [sic], J. W. Taylor, Lieut.

[P. S.] Our crew consisted of 31 men and boys, besides the officers; out of this number there were six seamen, the balance not knowing one part of the ship from the other, and it was with great difficulty that we obtained this crew while in New Orleans. (quoted in Dienst 1909)

Patrick O'Brian has written an acclaimed series of twenty novels about the British Royal Navy in the early 1800s, based on battle reports from the British Admiralty. The similarity of language and technical terms to the battle report above is striking, as is the postscript that most of the men did not know "one part of the ship from the other."

was platted a few miles upstream, but still on the tidal portion of the river. The new port was opened by the U.S. secretary of the treasury in 1891, and within a year more than one million dollars worth of lots were sold. The town had rail and riverboat connections to other cities and an area designated as Riverside Park. There was speculation that the English Rothschild family was involved in the town's development. When the 1900 hurricane hit, new Velasco had three thousand residents, and, of course, the town was destroyed (Weir 2009b).

Neither old nor new Velasco exists as a distinct community under that name today. New Velasco was annexed into Freeport in 1957. The site of old Velasco is now Surfside Beach, a community of beach houses and a county park. The U.S. Coast Guard Freeport Lifeboat Station is also at the site of old Velasco, where it has been since 1887. The station is on the spot where Austin's schooner *Lively* landed in 1821. A granite monument commemorates the event (Weir 2009b; U.S. Coast Guard 2009).

Brazosport is a "multicity" community made up of Brazoria, Clute, Freeport, Jones Creek, Lake Jackson, Oyster Creek, Quintana, Richwood, and Surfside Beach. Brazosport is primarily a chamber of commerce with the mission to "provide unified support and promotion of the Brazosport Community for economic

prosperity" (Brazosport Area Chamber of Commerce 2009). However, the name is now used for the school district that serves most of the area, for the community college, and for a regional health care system. Many of the residents live in unincorporated places outside the limits of formal towns, so Brazosport functions as an informal community even though it is not an actual incorporated unit of government (Kleiner 2009b).

SUGAR, COTTON, RICE, AND CARPETGRASS

The Indians of the lower Brazos did not farm and the Spaniards barely passed through, but the Anglo settlers viewed the abundant rainfall, rich alluvial soil, and long growing season offered by the lower Brazos valley as a major source of wealth. Stephen F. Austin sought out the best land, but he knew there had to be transportation as well. Thus, his settlement focused on the Brazos River. Sugar and cotton were the first cash crops, replaced later by rice, and now by subdivisions with carpetgrass.

Sugar

Christopher Columbus brought sugarcane to the New World on his second voyage and established a plantation and sugar mill on the island of Santo Domingo. Plantations and mills were developed in the Caribbean islands, South America, and finally the southern part of North America. Sugar production was

(and is) hot, hard, nasty work because the cane stalks had to be burned and then cut with cane knives. The stalks were crushed and the liquid evaporated in a series of copper or lead vats, called a "train," heated by wood fires to yield sugar crystals and molasses (Few 2007, 1).

Between 1850 and 1860 there were forty-six plantations in Brazoria County. Nineteen of those produced sugar, sixteen produced cotton, and three produced both sugar and cotton. Because sugar production is labor intensive and the work is so unpleasant, plantation owners obtained slaves to provide the labor, which established the initial demand for African slaves in the New World. Many of Austin's colonists brought slaves with them and received a bonus for each slave. Galveston and Houston had active slave markets, with dealers who kept slaves on hand for immediate delivery. The slave population in Brazoria County was three times the size of the white population in 1860 (Kleiner 2009a).

Emancipation of the slaves did not end the sugar business in Texas or the exploitation of workers, especially African Americans. Beginning in 1867 the state legislature allowed the prison population to be leased for labor. This policy finally ended in 1912. The owners of the twelve-thousand-acre Sugar Land plantation leased a large number of convicts, of which about 60 percent were black. Sugar Land came to be known as the

Varner-Hogg Plantation State Historic Site provides insight into the mechanics of sugar production from cane, but we cannot begin to understand the hard work that was done by slaves and later by convicts.

*Go down Old Hannah, don'cha rise
  no more
Don't you rise up til Judgment Day's
  for sure*

*Ain't no more cane on the Brazos
It's all been ground down to molasses*

*Captain, don't you do me like you done
  poor old Shine
Well ya drove that bully 'til he went
  stone blind*

*Wake up on a lifetime, hold up your
  own head
Well you may get a pardon and then
  you might drop dead*

*Ain't no more cane on the Brazos
It's all been ground down to molasses.*

"Hell Hole on the Brazos" (Texas Beyond History 2009).

The prison work song "Ain't No More Cane on the Brazos" expressed the feelings of convicts working on the sugar plantations:

*You shoulda been on the river in 1910
They were driving the women just like
  they drove the men.*

Domestic sugar production was protected by a high federal tax on imported sugar, which ended in 1928. About that time disease began to infect the cane, so the sugar refinery turned to imported raw sugar. There ain't no more sugarcane on the Brazos.

Cotton

People have made cloth from cotton for thousands of years, in both the Old World and the New. The Moors brought cotton to Spain, and the Spanish brought their own cotton to the New World. However, the people of the New World had also been using cotton for thousands of years. Today Mexican cotton (*Gossypium hirsutum*) comprises more than 90 percent of the cotton planted in the world.

Until it was mechanized in the 1950s cotton production was labor intensive like sugar and was most profitable using slave labor. The Brazos valley was a major cotton-producing area, especially the counties farther up the river where the growing season was too short for sugarcane. After the Civil War convict labor was used to produce cotton, and the state established an extensive system of prison farms ultimately totaling about one hundred thousand acres. More than forty-three thousand acres of prison farms were in the lower Brazos valley, on land that was previously worked by slaves (Lucko 2009).

Cotton production fell drastically in the 1920s due to the boll weevil and the pink boll worm, and it continued to decline as a result of competition from foreign producers and synthetic fibers. After World War II the majority of cotton production shifted to irrigated fields on the Texas High Plains. However, mechanization, improved farming methods, and better pest control have made it possible for farmers in the lower Brazos valley to continue to produce cotton (Britton, Elliott, and Miller 2008).

Rice

The ranching stereotype of Texas is familiar, but we probably do not think of rice paddies in the state. Yet beginning in the early 1900s rice production has been important on those rich, level, well-watered prairies of much of the Texas coast, including the lower Brazos. Rice was introduced to the Carolinas in the late 1600s from Madagascar but did not become important in Texas until 1904, when the Houston Chamber of Commerce and the Southern Pacific Railroad invited Seito Saibara from Japan to advise Gulf Coast farmers in rice production (Dethloff, 2009). Rice needs copious amounts of water, and the rivers of the state were the obvious source.

In 1902 Thomas U. Taylor, professor of civil engineering at the University of Texas at Austin, published "Rice Irrigation in Texas" in the *Bulletin of the University of Texas*. Writing of the Brazos and its tributaries he noted that "their joint flow below old San Felipe will have to be husbanded and stored if the canals now in existence and those now being constructed receive sufficient water for their rice" (Taylor 1902, 22). In 1910 the U.S. Department of Agriculture published a report titled *Irrigation in Texas* by C. Nagle, professor of civil engineering at the Agricultural and Mechanical College of Texas. Professor Nagle wrote that "in Fort Bend and Brazoria counties considerable water is now being pumped from the Brazos for rice irrigation, but opportunities still exist for large development along this and other lines in these counties" (Nagle 1910, 20). Regarding the Cane and Rice Belt Irrigation Company canal that took water from the Brazos River about twenty miles above Richmond, he wrote that "the opportunities for settlement are good, and there is room under the canal for enough to take up 8,000 acres, Germans or good Americans preferred" (Nagle 1910, 43).

Think what we might of Professor Nagle's ethnic recommendations, he and Taylor were both right about the availability of water from the Brazos for rice irrigation, but it depended on the husbanding of the entire system, as Taylor realized.

An important environmental benefit comes from the flooded rice paddies: they function somewhat like the natural wetlands, of which almost 50 percent on the Texas coast have been destroyed over the past fifty years (U.S. Fish and Wildlife Service 2009). Migratory birds of many species feed on the rice, including several species of geese that are popular game birds (Cockrell 2005).

Turfgrass

Neither Taylor nor Nagle could possibly have visualized what happened in the one hundred years after they made their recommendations. In

It is fitting that Texas A&M University, the state's land grant agricultural research university, should be located in a region that is so abundantly productive. As with agriculture in the other parts of Texas, an occasional oil well improves cash flow.

1900 Texas had about three million residents, and almost all of them made their living from agriculture. Today the state has twenty-four million people and less than 2 percent work in agriculture. In 1900 Harris County had sixty-four thousand residents; today it has about four million. Brazoria County had a population of fifteen thousand in 1900 and about three hundred thousand today. Oil had just been discovered and the Ford Model T, the car for the masses, was introduced in 1908, so most people still used horses for transportation. The petrochemical industry was in its infancy. Thus, in the context of the early twentieth century the Brazos was a crucial agricultural resource, which would lead Nagle to write, "Through the Trinity, Brazos, Colorado, Guadalupe, and Rio Grande, 20,000,000 acre-feet of water annually flows to the Gulf, which is lost so far as Texas land is concerned" (Nagle 1910, 88). In other words, water that made it to the mouth of the river was wasted. As we noted earlier, this sentiment persisted well into the twentieth century, and perhaps into the twenty-first, in the policies that govern development and management of the Brazos and other rivers in Texas.

The context for the Brazos River today is incomparably different from a hundred years ago. Fly to Sugar

The Brazos continues to provide water for rice production.

Land on Google Earth®. There at the site of one of the largest old plantations, note the drainage ditches that move storm water quickly into the river and the levees that constrain the river. But you won't see sugar or cotton within those drained and protected areas. What you see are thousands of acres of subdivisions with street names such as Misty Morn Lane and Gaelic Hill Lane. There is some turfgrass, but not much because the lots are so small and the houses so large there is little space left for lawn and much of that may contain a swimming pool. The golf courses are manicured turfgrass. The Brazos supplies some of the water for these communities.

This part of Texas is often perceived as being overdeveloped and polluted. In *The Historic Seacoast of Texas* David McComb expressed dismay about the area:

> For the visitor today, the distant past is hardly visible, and there is little connection to the present. Surrounding the old site of Quintana, for example, are the tanks and silver towers of the petrochemical industry.
>
> The employees in the petrochemical plants are migrants focused upon their work. They are not bad people, just rootless, with little connection to the lower Brazos. They will not linger beyond the time when the resources are gone: there is no expressed love of the land; the past has been brushed aside. (in Salvant 1999, 40)

I do not agree with McComb. The past is not very visible on the coast because the people of the past generally built flimsy structures that were frequently destroyed by hurricanes. People there do express love for the land and the water. While there are many suburbs consisting of large houses on small lots, there are also many pleasant homes on large lots in semirural areas, surrounded by colorful flowerbeds and sheltered by the low-hanging limbs of oaks. There is almost invariably a fishing boat nearby, indicating that the owners are attracted to the water. I've talked to people at the Brazos boat ramp under the Highway 36 bridge. They know the river and love fishing there. The rootlessness that McComb describes is a characteristic of the modern world, not Brazoria County alone. If anything, I would say there is more evidence of being rooted in Brazosport than in the wealthy suburbs of Austin or Dallas.

The slaves and convicts that slashed sugarcane on this Brazos land knew it as the hell hole of Texas, but today it provides beautiful homes for hundreds of thousands of people. Probably these are happy homes, with golf courses and parks, so things are better on the lower Brazos, at least for people. But what about the river itself? The theme of this chapter is that the modern lower Brazos is not pristine, but almost natural. Communities such as Sugar Land and Lake Jackson are recognizing the value of nature and taking steps both to preserve riparian areas and to make them accessible to people. This recognition puts the Brazos in a different context, one of honor and respect rather than exploitation. We are at a turning point and are now making decisions that will determine our relationship with the Brazos in the future, probably for much longer than the one hundred years that have passed since Nagle and Taylor made their recommendations. Can we see better into the future than they did? Probably not. But we can learn from the past four hundred years that Europeans have been on the Brazos and hopefully use that wisdom to direct our decisions. The final chapter of this book, "The Evolving Brazos," is an effort to alert us and perhaps inspire us to make careful decisions.

Where to Experience the Land and the River

The following locations are arranged in order from upstream to downstream.

WASHINGTON-ON-THE-BRAZOS STATE HISTORIC SITE: BOX 305, WASHINGTON, TEXAS 77880-0305
936-878-2214

Today Old Washington is very quiet. The park is for day use only, and most of the visitors seem quietly respectful of this place that is so important

to the history of Texas, as described earlier in this chapter. The Star of the Republic Museum, reconstructed buildings, and the Barrington Living History Farm help transport you back to the mid-1800s. When you stand on the bank looking down at the Brazos River, letting the quiet settle in, you begin to sense what life was like on the Brazos for immigrants almost two hundred years ago. You can look upstream and see the mouth of the Navasota where it joins the Brazos. You can look eastward and think about the flood of 1842 that Dr. Lockhart described as extending for miles through the forest and into the prairie. You can imagine people's excitement when they heard the whistle of an approaching steamboat, linking them in a tenuous way to the world they came from. At its peak population Washington had about 250 residents and was an important seat of government and commerce, but the town slowly died after its leaders kept faith in steamboats on the Brazos and refused the railroad.

STEPHEN F. AUSTIN STATE PARK: PARK ROAD 38, P.O. BOX 125, SAN FELIPE, TEXAS 77473-0125
979-885-3613

Most of the original town site of San Felipe is now within the Stephen F. Austin State Park. This park does not have the impressive interpretive facilities found at Washington-on-the-Bra-

zos, nor does it offer the opportunity to see alligators as easily as in Brazos Bend State Park. However, it is one of our favorite places on the lower Brazos. There are several trails that provide beautiful views of the river and somewhat difficult access. The trails wind through the tall bottomland forest, where we have seen pileated woodpeckers. I sit on the bluff looking over the river, amazed at the sense of wilderness, knowing that Houston is just a few miles away and a spot such as this is better than a museum. I use my imagination to think what life was like here when Houston was a controversial military leader, not a city.

BRAZOS BEND STATE PARK: 21901 FM 762, NEEDVILLE, TEXAS 77461
979-553-5102

Brazos Bend State Park is a treasure house of nature, from alligators to astronomy. The five-thousand-acre park is only twenty miles from Houston yet provides opportunities to experience and understand the richness and complexity of riparian forests, wetlands, and coastal prairies. There are more than three hundred bird species, twenty-three mammal species, and twenty-one reptile spe-

Stephen F. Austin State Park is one of the reasons that the (Almost) Free Brazos is so important—because it provides access to wild nature only a short drive from Houston.

CAUTION-ALLIGATORS
EXIST IN THIS PARK
· DO NOT FEED OR ANNOY ·
STAY AT A DISTANCE

cies in the park. Of course, the most exciting of the reptiles are the three hundred American alligators that live in the park. You can stand on docks overlooking oxbow lakes and sloughs and safely watch alligators. Park officials are very clear that alligators are dangerous wild animals and will harm people if we disturb them. But that is the excitement of Brazos Bend—these are not robotic replicas or captive animals in a zoo, these are the real thing in the wild.

The Houston Museum of Natural Science operates the George Observatory and Challenger Learning Center at the park, helping visitors experience the immensity of space while possibly hearing the mating call of a bull alligator.

Brazos Bend State Park, with its campgrounds and trails, is a place where people can have a lifetime of learning and enjoyment. More than 80 percent of Texans live in urban areas, so a Texas child will be astonished and awed by the alligators, deer, raccoons, and other animals in the wild. Access to the Brazos River is provided, so those seeking the adventure of a remote river trip can have the opportunity. The park's location

Although direct access to the Brazos is somewhat limited at Brazos Bend State Park, the old oxbow lakes left by the river offer outstanding aquatic habitats for creatures from the sublime to the awesome.

on one of the major migratory bird routes and its variety of habitats will challenge a birder for years. There are dragonflies, butterflies, grasses, trees, wildflowers—opportunities for learning, photography, and painting. And then there is the universe seen through the telescopes. We do not have to travel to Africa or the Amazon, not to mention a theme park, to have a high quality adventure. The Brazos River and its Big Creek tributary provide those adventures within minutes of the largest city in the state.

VARNER-HOGG PLANTATION STATE HISTORIC SITE: 1702 N. THIRTEENTH STREET, WEST COLUMBIA, TEXAS 77486 979-345-4656

The Varner-Hogg Plantation State Historic Site embraces the drama of the first one hundred years of Anglo Texas history. It was one of the earliest plantations in Austin's Colony, perhaps producing the first rum in Texas. The physical work was done by slaves, and one owner had a slave mistress who kept her own slaves. Slaves built a magnificent house using bricks made with clay from the banks of the Brazos. The plantation experienced wealth and poverty and became a second home for Governor James Hogg. Finally, to complete the Texas story, oil was discovered on the property. The plantation house and several outbuildings are in outstanding condition.

SEA CENTER TEXAS: 300 MEDICAL DRIVE, LAKE JACKSON, TEXAS 775636 979-292-0100

Sea Center Texas is a marine aquarium, fish hatchery, and nature center with exhibits and aquaria that demonstrate native aquatic habitats, including a salt marsh, jetty, reef, and open Gulf exhibits. Although these are not riverine habitats, they do depend to some extent on the outflow from the Brazos and other rivers.

CHAPTER 7
The Evolving Brazos

The ultimate goal of any renewal process must be to establish a mutually enhancing mode of human-Earth relations. . . . This story of the universe now becomes the basic framework for education. . . . If we respond to it properly, this story can guide us through the transition phase of our history from the terminal Cenozoic into the emerging Ecozoic. This new, emergent phase of Earth history can be defined as that period when humans would be present to the Earth in a mutually enhancing manner.

Thomas Berry, *Evening Thoughts* (2006)

THOMAS BERRY was a cultural historian and a priest who called himself a "geologian." He contended that humans have so extensively modified Earth's biophysical systems that we have brought an end to the Cenozoic era and have created the "Ecozoic" era, in which we are the dominant force. We are in charge, for better or worse.

In the Brazos River context, do the previous chapters support Berry's claim? Perhaps only partially. The Brazos is Texas' most regulated (dammed) river, but most of its upper reaches still run free and the lower portion is not dammed and has not been radically affected by the dams in the central part of the basin. But evolution is an ongoing process, and our species is intelligent and aggressive. We are not through yet. The big question is, what will we do?

The most immediate change to be made is the off-stream Allen's Creek project on the lower part of the river. The Texas Commission on Environmental Quality has issued permits to dam Allen's Creek, a tributary of the Brazos near Simonton, Texas, and pump water from the Brazos into the reservoir. Houston and other Gulf Coast communities will use this water. The project is permitted to with-

How many thousands of years have children and their dogs been attracted to the Brazos River?

draw a maximum of 202,000 acre-feet per year from the Brazos (Osting, Mathews, and Austin 2004), and it cannot pump more than a set amount per month, which varies according to season. The highest monthly amount allowed is 2,200 cfs, and pumping is not allowed to reduce flow below the standard water quality protection flow level (734 cfs).

As noted in chapter 5, the topography of the lower part of the Brazos is not well suited for dams because there are few natural low places that can retain water. The Allen's Creek site consists of a bluff on one side of the creek. A dam more than four miles long and about fifty feet high will be built on the other side of the creek, so basically the Brazos River Authority and the City of Houston are creating the necessary topographic depression. The reservoir will cover about ninety-five hundred acres and yield an estimated one hundred thousand acre-feet per year. The City of Houston and the authority will sell the water to various users (Brazos River Authority 2002).

The Brazos River Authority has operated under a "system order" since 1964, which allows it to release water from the various reservoirs in the system as the water is demanded by the downstream purchasers. The system order results in efficiencies that in effect increase the yield of the system. However, the existing system order does not give the authority rights to that additional water. Thus, the Brazos River Authority is currently requesting an additional water right of 421,449 acre-feet of firm yield plus 670,000 acre-feet of interruptible yield (which would require commitment of 90,000 acre-feet of its firm yield), for a total new water right of 1,001,449 acre-feet. In addition, until the Allen's Creek project is completed, the authority is requesting rights for 1,204,009 acre-feet, of which 425,099 would be firm yield. The proposal was submitted in 2004 and has been challenged by a vari-

ety of parties, including water rights holders and environmental interests. As of 2010 the proposal is going through the formal hearings process within the Texas Commission for Environmental Quality and the State Office of Administrative Hearings.

Even though the Brazos is muddy and still somewhat salty by the time it reaches its lower course, the Texas Gulf Coast region is rapidly growing and the Brazos can help meet its demand for water. Thus, there are strong economic and political forces that will work to wring every bit of water possible out of the Brazos. The Brazos can be greatly harmed as an ecological system unless the Brazos River Authority and other agencies and users base their policies and actions on scientific knowledge about the essential volume and timing of flows in the Brazos.

But there are also strong political, economic, and cultural/psychological forces for conservation that counter or perhaps at least balance the forces for exploitation. Many people understand Thoreau's statement "in wildness is the preservation of the world" (Thoreau n.d., loc. 215), and it appears that more are beginning to realize the importance of nature, even on the Brazos. The Brazos between Possum Kingdom Lake and Waco

The Falls of the Brazos at Marlin and other similar shoals prevented Waco from becoming an inland port. On a bright winter day they remind us of our inextricable connection to nature.

has become one of the most popular canoeing streams in Texas. Yes, it is popular because it is fairly clear due to the dams, but at least people are out there.

We believe that with better knowledge and understanding of the Brazos an increasing number of people will recognize the value of nature represented in the wilder parts of the river, from the salty tributaries to the wide, muddy lower river. Previous chapters described places where people can enjoy those qualities, but those places are insufficient. Thus, in our Ecozoic evolutionary process those of us who recognize the value of wildness must speak loudly to retain wild places and to provide access to them.

We have looked back into the past of the Brazos River in both geologic time and human history. It is a complex story. More importantly we need to look into the future of the Brazos because we are shaping its future by the decisions we make now. What do we want from the Brazos? Do we want it to be a conveyance that carries water to thirsty cities and industries? Do we want it to be a wild place that instructs, inspires, and maybe even frightens us? Yes, we want it to be all of that and more. Other than climate we humans are the main deciding factor for the future of the Brazos River.

John Graves described this part of the Brazos at the mouth of the Paluxy in Goodbye to a River, *his protest of the Six Dam Plan that would have inundated much of the central Brazos. Since 1960 Graves's book has inspired Texans about part of the Brazos. Now we need to* abrazar *(embrace) the entire Brazos de Dios.*

Chapter 7

APPENDIX

Plant and Animal Species of the Brazos River

Riparian Vegetation

The riparian corridor is the area immediately adjacent to the river, where the soil has high moisture content because it is in contact with water from the river. It is a hospitable place for many plants, but they must be ones that need or tolerate high moisture and periodic flooding. Riparian vegetation provides rich habitat, furnishes organic matter to the river, protects the banks from erosion, and retards floodwaters. The riparian corridor is narrow in the upper tributaries and is absent where the river has cut steep banks.

Common Riparian Vegetation of the Upper and Middle Brazos

baccharis (*Baccharis* spp.)
cottonwood (*Populus* spp.)
elm (*Ulmus* spp.)
hackberry (*Celtis* spp.)
mesquite (*Prosopis* spp.)
salt cedar (*Tamarix* spp.)
   (non-native)
sumac (*Rhus* spp.)
sycamore (*Platanus occidentalis*)
willow (*Salix* spp.)

Sources: **Dahm, Edwards, and Gelwick 2005, 206; Vines 1984.**

The hardwood bottomland forests of the lower Brazos, San Bernard, and Colorado rivers are extremely rich ecosystems with more than three hundred species of plants (Rosen, De Steven, and Lange 2008, 77). The table below lists the most common species and also illustrates how sensitive vegetation is to the slight elevation differences between the ridges, flats, and meander scars or channels.

Vegetation of the Lower Brazos Bottomlands

OVERSTORY/UNDERSTORY

Species	Average relative importance value (%)		
	Ridge	Flat	Meander scar
Laurel cherry (*Prunus caroliniana*)	40	6	-
Sugarberry (*Celtis laevigata*)	10	8	<1
Water oak (*Quercus nigra*)	10	2	<1
Soapberry (*Sapindus saponaria*)	5	9	-
Live oak (*Quercus virginiana*)	-	12	-
Cedar elm (*Ulmus crassifolia*)	3	16	<1
Yaupon (*Ilex vomitoria*)	14	23	-
Green ash (*Fraxinus pennsylvanica*)	3	2	30
Swamp-privet (*Forestiera acuminata*)	-	-	28
Buttonbush (*Cephalanthus occidentalis*)	-	-	16
Water hickory (*Carya aquatic*)	-	-	5
Chinese tallowtree (*Triadica sebifera*)	-	-	6

GROUND LAYER

Species	Average cover (%)		
	Ridge	Flat	Meander scar
Laurel cherry (*Prunus caroliniana*)	9	1	-
Basket-grass (*Oplismenus hirtellus*)	6	9	-
Virginia jumpseed (*Tovara virginiana*)	2	9	-
Hairy-collar wood-oats (*Chasmanthium laxum*)	1	10	-
Poison ivy (*Toxicodendron radicans*)	1	15	<1
Cherokee sedge (*Carax cherokeensis*)	<1	30	-
Dwarf palmetto (*Sabal minor*)	4	23	11
Little duckweed (*Lemna obscura*)	-	-	29
Heart-leaf burhead (*Echinodorus cordifolius*)	-	-	18
Swamp panic-grass (*Penicum gymnocarpon*)	-	-	14
Gulf swampweed (*Hygrophilia lacustris*)	-	-	8

Source: Adapted from Rosen, De Steven, Lange 2008, 81.

Aquatic Plants

Aquatic plants have a great variety of forms due to the diversity of habitats in a river. Some are free floating (phytoplankton), some attach themselves to rocks, sticks, or other solid substrate (biofilm or periphyton), some are rooted in the bottom and remain submerged under the water, while other rooted plants extend above the water surface.

AQUATIC PLANTS
IN THE BRAZOS
(adapted from Dahm, Edwards, and Gelwick 2005, 206)

Algae
   diatoms: *Nitzchia, Navicula, Cymbella, Gomphonema, Diatoma, Synedra, Navicula, Tabellaria, Cocconema, Cosmarium*
   unicellular green algae: *Ankistrodesmus, Characium*
   filamentous green algae: *Rhizodonium, Cladophora, Oedogonium, Spirogyra, Tribonema, Mougeoutia, Ulothrix*
   Cyanobacteria: *Anabaena, Oscillatoria*
   Golden algae: *Prymnesium parvum* (toxic exotic)

(adapted from Dahm, Edwards, and
Gelwick 2005, 206)

Native

arrowhead (*Sagittaria* spp.)

duckweed (*Lemna* spp.)

eelgrass (*Zostera* spp.)

pondweed, Illinois (*Potamogeton illinoensis*)

primrose, water (*Ludwigia peploides*)

spatterdock (*Nuphar polysepala*)

spikerush (*Eleocharis palustris*)

stargrass, water (*Zosterella dubia*)

water shield (*Brasenia schreberi*)

willow, water (*Justica* spp.)

Non-native

alligatorweed (*Alternanthera philoxeroides*)

duckweed, dotted (*Landoltia punctata*)

hyacinth, water (*Eichhornia crassipes*)

hydrilla (*Hydrilla verticillata*)

salvinia, giant (*Salvinia auriculata*)

water lettuce (*Pistia stratiotes*)

Invertebrates

Most of the life forms you will see in the Brazos or any other body of water are the invertebrates, which include insects, crustaceans, mollusks, and oligochaetes. In North America there are about ten thousand aquatic insect species. The most common of these are the Diptera (true flies), Ephemeroptera (mayflies), Plecoptera (stoneflies), and Trichoptera (caddisflies) (Benke and Cushing 2005, 11–12). As you can see in the lists below, the Brazos has a large variety of invertebrates. The Leon River winter stonefly (*Taeniopteryx starki*) is a threatened or endangered aquatic insect in the Brazos system.

MOST COMMON INVERTE-
BRATES IN THE BRAZOS RIVER
(adapted from Dahm, Edwards, and
Gelwick 2005, 206, 223)

Some invertebrates are sensitive to water conditions, so the species composition of the invertebrate community varies along the various reaches of the Brazos:

Upper Brazos

beetles:

Stenelmis sexlineata

Stenelmis occidentalis

Berosus subsignatus

biting midges: *Bezzia* spp.

caddisflies: *Cheumatopsyche* spp.

chironomid midges:

Polypedilum scalaenum

Polypedilum convictum

Rheotanytarsus exiguous

Cryptochironomus fluvus

oligochaete worms: *Limnodrilus hoffmeisteri*

Middle Brazos

beetles: *Psephenus texanus*

caddisflies (most common out of more than forty species):

Hydropsyche simulans

Cheumatopsyche lasia

Cheumatopsyche campyla

Chimarra obscura

Hydroptila spp.

Oxyethira spp.

chironomids:

Orthocladius spp.

Cricotopus bicinctus

Tanytarsus glabrescens

Tanytarsus guerlus

Dicrotendipes neomodestus

Pseudochironomus spp.

damselflies: *Argia* spp.

mayflies:

Stenonema spp.

Chroterpes mexicanus

Fallceon quilleri

Thraulodes gonzalesi

Tricorythodes albilineatus

stoneflies:

Perlesta placida

Neoperla clymene

Brazos below Waco

beetles: *Stenelmis occidentalis*

caddisflies:

Smicridea spp.

Cheumatopsyche spp.

chironomids:

Pentaneura spp.

Meropelopia spp.

dragonflies:

Brechmorhoga mendax

Erpetogomphus spp.

mayflies:
  Fallceon quilleri
  Baetis pygmaeus
oligochaete worms:
  Dero digitata
  Limnodrilus maumeensis
  Limnodrilus hoffmeisteri
  Quistadrilus multisetosus
phantom midges: *Chaoborus* spp.
snails: creeping ancylid (*Ferrissia* spp.)
tipulid flies: *Hexatoma* spp.

Crayfish
swamp dwarf crayfish:
  Cambarellus puer
  Cambarellus texanus
  Fallicambarus hedgpethi
  Orconectes causeyi
  Orconectes palmeri longimanus
white crayfish: Orconectes virilis
White River crayfish:
  Procambarus acutus acutus
red swampy: *Procambarus clarkia*

Mussels
(most common of nineteen species)
Asian clam (*Corbicula fluminea*) (non-native)
Bleufer (*Potamilus* spp.)
false spike (*Elliptio* spp.)
golden orb (*Quadrula aurea*)
pimpleback (*Quadrula* spp.)
pistolgrip (*Tritogonia verrucosa*)
pondhorn (*Uniomerus* spp.)
southern mapleleaf (*Quadrula* spp.)
Texas fawnsfoot (*Truncilla* spp.)
Texas lilliput (*Toxolasma* spp.)

Brazos River Fish

The great diversity of habitats in the long and complex Brazos provides opportunities for ninety-two fish species. Threatened or endangered fish in the Brazos system include the blue sucker (*Cycleptus elongates*), Guadalupe bass (*Micropterus treculii*), sharpnose shiner (*Notropis oxyrhynchus*), and the smalleye shiner (*Notropis buccula*) (HDR Inc. 2001, E-5).

FISH IN THE BRAZOS RIVER
(most common of the ninety-two species; adapted from Dahm, Edwards, and Gelwick 2005, 207, 223)

Native
bass, largemouth (*Micropterus salmoides*)
bass, spotted (*Micropterus punctulatus*)
bluegill (*Lepomis macrochirus*)
buffalo, smallmouth (*Ictiobus bubalus*)
catfish, black bullhead (*Ameiurus melas*)
catfish, channel (*Ictalurus punctatus*)
catfish, flathead (*Pylodictis olivaris*)
catfish, yellow bullhead (*Ameiurus natalis*)
crappie, white (*Pomoxis annularis*)
darter, dusky (*Percina sciera*)
darter, orangethroat (*Etheostoma spectabile*)
darter, slough (*Etheostoma gracile*)
gar, longnose (*Lepisosteus osseus*)
gar, spotted (*Lepisosteus occulatus*)
killifish, plains (*Fundulus zebrinus*)
madtom, tadpole (*Noturus gyrinus*)
minnow, bullhead (*Pimephales vigilax*)
minnow, central stoneroller (*Campostoma anomalum*)
minnow, plains (*Hybognathus placitus*)
mosquitofish (*Gambusia* spp.)
perch, pirate (*Aphredoderus sayanus*)
pupfish, Red River (*Cyprinodon rubrofluviatilis*)
shad, gizzard (*Dorosoma cepedianum*)
shiner, blacktail (*Cyprinella venusta*)
shiner, golden (*Notemigonus crysoleucas*)
shiner, mimic (*Notropis volucellus*)
shiner, red (*Cyprinella lutrensis*)
shiner, sharpnose (*Notropis oxyrhynchus*)
shiner, smalleye (*Notropis buccula*)
sucker, spotted (*Minytrema melanops*)
sunfish, banded pygmy (*Elassoma zonatum*)
sunfish, green (*Lepomis cyanellus*)

sunfish, orangespotted (*Lepomis humilis*)

sunfish, redear (*Lepomis microlophus*)

topminnow, blackstripe (*Fundulus notatus*)

warmouth (*Lepomis gulosus*)

Nonnative

bass, smallmouth (*Micropterus dolomieu*)

bass, striped (*Morone saxatilis*)

carp, common (*Cyprinus carpio*)

cichlid, Rio Grande (*Cichlasoma cyanoguttatum*)

pike, northern (*Esox lucius*)

rudd (*Scardinius erythrophthalmus*)

sauger (*Stizostedion canadense*)

silverside, inland (*Menidia beryllina*)

tilapia, blue (*Oreochromis aureus*)

trout, rainbow (*Oncorhynchus mykiss*)

walleye (*Stizostedion vitreum*)

Reptiles of the Brazos River

All of the snake species below are adapted to live in water. Although not aquatic species, the copperhead (*Agkistrodon contortrix*) and timber rattlesnake (*Crotalus horridus*) are found in riparian areas and floodplain forests. Threatened or endangered reptiles in the Brazos system include the alligator snapping turtle (*Macrochelys temminckii*), Brazos (Harter's) water snake (*Nerodia harteri*), and

the Concho water snake (*Nerodia paucimaculata*).

COMMON BRAZOS RIVER
REPTILES
(adapted from Dahm, Edwards, and Gelwick 2005, 207)

American alligator (*Alligator mississippiensis*)

blotched water snake (*Nerodia erythrogaster transversa*)

broad-banded water snake (*Nerodia fasciata confluens*)

Concho water snake (*Nerodia paucimaculata*)

cottonmouth (*Agkistrodon piscivorous*)

diamondback water snake (*Nerodia rhombifer*)

glossy crayfish snake (*Regina rigida*)

green water snake (*Nerodia cyclopion*)

Harter's water snake (*Nerodia harteri*)

Ouachita map turtle (*Graptemys ouachitensis*)

red-eared slider (*Trachemys scripta elegans*)

smooth softshell turtle (*Apalone mutica*)

spiny softshell turtle (*Apalone spinifera*)

stinkpot turtle (*Sternotherus odoratus*)

Texas map turtle (*Graptemys versa*)

Amphibians of the Brazos River

The complex amphibious natural history of these animals makes them especially vulnerable to environmental changes. Worldwide, amphibians are going extinct at a rate two hundred times faster than they have over the past 350 million years. Population sizes are declining in at least 43 percent of the species (Collins, Gascon, and Mendelson 2007). Brazos amphibians that are threatened or endangered are the Georgetown salamander (*Eurycea naufragia*), the Houston toad (*Bufo houstonensis*), the Jolleyville plateau salamander (*Eurycea tonkawae*), and the Salado Springs salamander (*Eurycea chisholmensis*) (HDR Inc. 2001, E-5, E-6).

COMMON BRAZOS RIVER
AMPHIBIANS
(adapted from Dahm, Edwards, and Gelwick 2005, 207)

cricket frog (*Acris crepitans*)

green frog (*Rana clamitans*)

plains leopard frog (*Rana blairi*)

Rio Grande leopard frog (*Rana berlandieri*)

southern leopard frog (*Rana sphenocephala utricularia*)

spotted chorus frog (*Pseudacris clarkii*)

Strecker's chorus frog (*Pseudacris streckeri streckeri*)

Brazos Birds and Mammals

Texas provides habitat for 615 species of birds, 333 of which are migratory (Texas Parks and Wildlife Department 2009). Given that the Brazos cuts across most of the ecoregions of Texas, most of these birds use some part of the Brazos River's water and riparian corridor. Texas is on the Central Flyway for migratory birds, and the Brazos River cuts directly across that flyway.

BIRDS

Threatened or endangered birds that rely on habitats provided by the Brazos include the following: arctic peregrine falcon (*Falco peregrinus tundrius*), bald eagle (*Haliaeetus leucocephalus*), ferruginous hawk (*Buteo regalis*), interior least tern (*Sterna antillarum athalassos*), migrant loggerhead shrike (*Lanius ludovicianus migrans*), piping plover (*Charadrius melodus*), snowy plover (*Charadrius alexandrinus*), white-faced ibis (*Plegadis chihi*), whooping crane (*Grus americana*), wood stork (*Mycteria americana*), and the zone-tailed hawk (*Buteo albonotatus*).

For detailed information about birds in Texas consult the Texas Parks and Wildlife Department (http://www.tpwd.state.tx.us/huntwild/wild/birding/index.phtml). For information on migratory species in Texas see http://www.tpwd.state.tx.us/huntwild/wild/birding/migration/.

MAMMALS

beaver (*Castor canadensis*)
nutria (*Myocastor coypus*) (nonnative)
raccoon (*Procyon lotor*)
swamp rabbit (*Sylvilagus aquaticus*)

REFERENCES

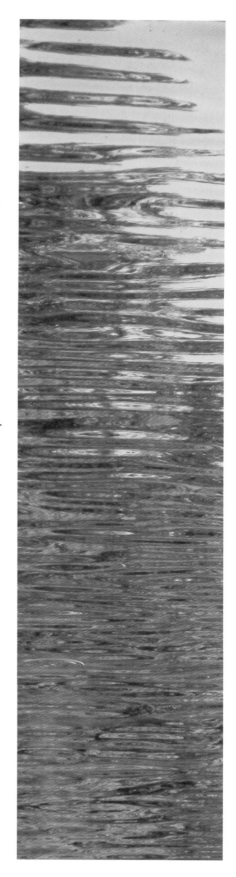

Chapter 1: Water Runs Downhill

Benke, Arthur C., and Colbert E. Cushing. 2005. Backgroun and Approach. *In Rivers of North America*, edited by Arthur C. Benke and Colbert E. Cushing, 1–18. Burlington, Mass.: Elsevier Academic Press.

Bomar, George W. 2010. Weather. *Handbook of Texas Online*. http://www.tshaonline.org/handbook/online/articles/yzw01 (accessed December 14, 2010).

Bourgeois, Joanne, Thor A. Hansen, Patricia L. Wiberg, and Erle G. Kauffman. 1988. A Tsunami Deposit at the Cretaceous-Tertiary Boundary in Texas. *Science*, n.s. 241(4865): 567–70.

Brazos G Regional Water Planning Group. 2006. *Brazos G Regional Water Planning Area. Regional Water Plan*. http://www.brazosgwater.org/fileadmin/brazosg_upload/2006_brazosg_water_plan/FINAL_Report_Materials/Volume%20I%20(all)%20(Updated%20April%202006).pdf

Brazos River Authority. 2007. *Clean Rivers Program Basin Summary Report*. http://www.brazos.org/BasinSummary_2007.asp (accessed November 14, 2007).

Burroughs, William J., Bob Crowder, Ted Robertson, Eleanor Vallier-Talbot, and Richard Whitaker. 1996. *A Guide to Weather*. San Francisco: Fog City Press.

Carlson, Paul H. 2005. *Deep Time and the Texas High Plains*: History and Geology. Lubbock: Texas Tech University Press.

Dunn, David D., and Timothy H. Raines. 2001. *Indications and Potential Sources of Change in Sand Transport in the Brazos River, Texas*. Water-Resources Investigations Report 01–4057. Austin, Tex.: U.S. Geological Survey.

Flannery, Tim. 2001. *The Eternal Frontier: An Ecological History of North America and Its Peoples*. New York: Grove Press.

Hendrickson, Kenneth E., Jr. 1981. *The Waters of the Brazos: A History of the Brazos River Authority, 1929–1979*. Waco: Texian Press.

Hentz, Tucker F. 2007. Geologic Timeline. *Handbook of Texas Online*. http://www.tsha.utexas.edu/handbook/online/articles/GG/swgqz.html (accessed September 3, 2007).

Holliday, Vance T. 1987. Cultural Chronology. *In Lubbock Lake: Late Quaternary Studies on the*

Southern High Plains, edited by
Eileen Johnson, 22–25. College
Station: Texas A&M University
Press.

Hudson, Paul F., and Joann Mossa.
1997. Suspended Sediment
Transport Effectiveness of Three
Large Impounded Rivers, U.S.
Gulf Coastal Plain. *Environmental Geology* 32(4): 263–73.

Leopold, Luna B. 1994. *A View of the River*. Cambridge, Mass.:
Harvard University Press.

McMahan, Craig A., Roy G. Frye,
and Kirby L. Brown. 1984. *The Vegetation Types of Texas including Cropland*. Austin: Texas
Parks and Wildlife Department.

Renfro, H. B. 1979. *Geological Highway Map of Texas*. Tulsa,
Okla.: American Association of
Petroleum Geologists.

Smith, Norman. 1971. *A History of Dams*. Secaucus, N.J.: Citadel
Press.

Spearing, Darwin. 1991. *Roadside Geology of Texas*. Missoula, Mt.:
Mountain Press Publishing.

U.S. Geological Survey. 2008.
Suspended-Sediment Database.
Daily Values of Suspended Sediment and Ancillary Data, Spatial
Patterns of Sediment Concentration in the United States. http://
co.water.usgs.gov/sediment/
conc.frame.html#HDR3 (July
19, 2008).

Wetzel, Robert G. 2001. *Limnology: Lake and River Ecosystems*. 3rd

ed. San Diego, Calif.: Academic
Press.

Winchester, Simon. 2004. *The River at the Center of the World*. New
York: Picador.

Chapter 2: The Brazos as an Ecological System

American Heritage Science Dictionary. http://dictionary.reference
.com/browse/ecosystem (accessed
June 19, 2009).

Benke, Arthur C., and Colbert
E. Cushing, eds. 2005. *Rivers of North America*. Burlington,
Mass.: Elsevier Academic Press.

Benke, Arthur C., Indrajeet
Chaubey, G. Milton Ward, and
E. Lloyd Dunn. 2000. Flood
Pulse Dynamics of an Unregulated River Floodplain in the
Southeastern U.S. Coastal Plain.
Ecology 81(10): 2730–41.

Fischer, Helmut, Frank Kloep, Sabine Wilzcek, and Martin Pusch.
2005. A River's Liver—Microbial
Processes within the Hyporheic
Zone of a Large Lowland River.
Biogeochemistry 76: 349–71.

Hendrickson, Kenneth E., Jr. 1981.
*The Waters of the Brazos: A History of the Brazos River Authority,
1929–1979*. Waco: Texian Press.

Postel, Sandra, and Brian Richter.
2003. *Rivers for Life: Managing Water for People and Nature*.
Washington, D.C.: Island Press.

Texas Commission on Environmental Quality, Texas Parks and
Wildlife Department, and Texas
Water Development Board. 2008.
Texas Instream Flow Studies: Technical Overview. Report 369.
Austin: Texas Water Development Board.

Texas Water Development Board.
2010. Brazos River Alluvium.
http://www.twdb.state.tx.us/publications/reports/GroundWater-
Reports/GWReports/R345%20
Aquifers%200f%20Texas/Minors/brazos.pdf

Thorp, J. H., and M. D. Delong.
1994. The Riverine Productivity Model: An Heuristic View
of Carbon Sources and Organic
Processing in Large River Ecosystems. *Oikos* 70: 305–8.

Vannote, R. L., G. W. Minshall,
K. W. Cummins, J. R. Sedell, and
C. E. Cushing. 1980. The River
Continuum Concept. Canadian
Journal of Fisheries and Aquatic Sciences 37: 130–37.

Voshell, J. Reese, Jr. 2002. *A Guide to Common Freshwater Invertebrates of North America*. Blacksburg, Va.: McDonald & Woodward Publishing.

Wetzel, Robert G. 2001. *Limnology: Lake and River Ecosystems*. 3rd
ed. San Diego, Calif.: Academic
Press.

Chapter 3: The Lost River

Blackburn, Elliott. 2010. Lake Meredith Continues to Sink to Record Lows. *Lubbock Avalanche-Journal*, January 14. http://www.lubbockonline.com/stories/011410/loc_548823829.shtml (March 23, 2010).

Bolen, Eric, Loren M. Smith, and Harold L. Schramm Jr. 1989. Playa Lakes: Prairie Wetlands of the Southern High Plains. *BioScience* 39(9): 615–23.

Bomar, George W. 1983. *Texas Weather*. Austin: University of Texas Press.

———. 2010. Weather. *Handbook of Texas Online*. http://www.tshaonline.org/handbook/online/articles/yzw01 (accessed December 14, 2010).

Bousman, C. Britt. 2004. Paleoindian Archeology in Texas. In *The Prehistory of Texas*, edited by Timothy K. Perttula, 15–97. College Station: Texas A&M University Press.

Canadian River Municipal Water Authority. 2008. http://www.crmwa.com/ (accessed June 19, 2008).

City of Lubbock 2008. Water Supply. http://water.ci.lubbock.tx.us/about/overview.htm (March 23, 2010).

Fite, Gilbert. 1977. Great Plains Farming: A Century of Change and Adjustment. *Agricultural History* 51(1): 244–56.

Fonstad, Mark, William Pugatch, and Brandon Vogt. 2003. Kansas Is Flatter Than a Pancake. *Annals of Improbable Research* 9(3).

Graves, Lawrence L. 2009. Lubbock, Texas. *Handbook of Texas Online*. http://www.tshaonline.org/handbook/online/articles/LL/hd14.html (accessed July 7, 2009).

Holliday, Vance T. 1987. Cultural Chronology. In *Lubbock Lake: Late Quaternary Studies on the Southern High Plains*, edited by Eileen Johnson, 22–25. College Station: Texas A&M University Press.

———. 1990. Soils and Landscape Evolution of Eolian Plains: The Southern High Plains of Texas and New Mexico. *Geomorphology* 3: 489–515.

Johnson, Eileen, and Vance T. Holliday. 1987. Introduction to *Lubbock Lake: Late Quaternary Studies on the Southern High Plains*, edited by Eileen Johnson, 3–13. College Station: Texas A&M University Press.

Jordan, Terry G. 1984. *Texas: A Geography*. With John L. Bean Jr. and William M. Holmes. Boulder, Colo.: Westview Press.

Kingsford, R. T., and J. R. Thompson. 2006. Desert or Dryland Rivers of the World: An Introduction. In *Ecology of Desert Rivers*, edited by Richard Kingsford, 3–10. Cambridge: Cambridge University Press.

Morris, John Miller. 1997. *El Llano Estacado*. Austin: Texas State Historical Association.

Ray, James D., Brian D. Sullivan, and Harvey W. Miller. 2003. Breeding Ducks and Their Habitats in the High Plains of Texas. *Southwestern Naturalist* 48(2): 241–48.

Sabin, Ty J., and Vance T. Holliday. 1995. Playas and Lunettes on the Southern High Plains: Morphometric and Spatial Relationships. *Annals of the Association of American Geographers* 85(2): 286–305.

Seyffert, Kenneth D. 2001. *Birds of the High Plains and Rolling Plains of Texas: A Field Checklist*. Austin: Texas Parks and Wildlife Department.

Spearing, Darwin. 1991. *Roadside Geology of Texas*. Missoula, Mt.: Mountain Press Publishing.

Sublette, James E., and Mary Smith Sublette. 1967. The Limnology of Playa Lakes on the Llano Estacado, New Mexico and Texas. *Southwestern Naturalist* 12(4): 369–406.

Sylvia, Dennis A., and William E. Galloway. 2006. Morphology and Stratigraphy of the Late Quaternary Lower Brazos Valley: Implications for Paleo-climate, Discharge, and Sediment Delivery. *Sedimentary Geology* 190: 159–75.

Tacha, T. C., S. A. Nesbitt, and P. A. Vohs. 1992. Sandhill Crane (*Grus canadensis*). *The Birds of North America Online*, edited by A. Poole. Ithaca, N.Y.: Cornell

Lab of Ornithology. http://bna
.birds.cornell.edu/bna/spe
cies/031. doi:10.2173/bna.31.

U.S. Census Bureau. 2009. State
and County QuickFacts, Lubbock
County, Texas. http://quickfacts
.census.gov/qfd/states/48/48303
.html (accessed July 10, 2009).

U.S. Department of Agriculture.
2009. Natural Resource Conservation Service Plants Data
Base. http://plants.usda.gov/java/
profile?symbol=DIACF (accessed
July 7, 2009).

U.S. Fish and Wildlife Service.
2010. Muleshoe National Wildlife
Refuge. http://www.fws.gov/
refuges/profiles/index.cfm?
id=21590 (accessed December
14, 2010).

Wester, David B. 2007. The Southern High Plains: A History of
Vegetation, 1540 to Present. In
*Proceedings: Shrubland Dynam-
ics—Fire and Water*, compiled
by R. E. Sosebee, D. B. Wester,
C. M. Britton, E. D. McArthur,
and S. G. Kitchen. August 10–12,
2004, Lubbock, Tex. Proceedings RMRS-P-47. Fort Collins,
Colo.: U.S. Department of Agriculture, Forest Service, Rocky
Mountain Research Station.

Chapter 4: Many Arms of God

Brazos River Authority. 2008. Special Studies and Activities. http://
www.brazos.org/crpPDF/05Ann
ualWaterHighlightsSpecial
Studies.pdf (July 14, 2009)

Carlisle, Jeffrey D. 2009. Apache Indians. *Handbook of Texas Online.*
http://www.tshaonline.org/hand
book/online/articles/AA/bma33
.html (accessed July 15, 2009).

Davis, Charles G. 2009. Crystal Falls,
Texas. *Handbook of Texas Online.*
http://www.tshaonline.org/hand
book/online/articles/CC/hncaq
.html (accessed August 24, 2009).

de la Teja, Jesús F., Paula Marks,
and Ron Tyler. 2004. *Texas:
Crossroads of North America.*
Boston: Houghton Mifflin.

Dudley, Tom L., and David J.
Kazmer. 2005. Field Assessment
of the Risk Posed by *Diorhabda
elongata*, a Biocontrol Agent for
Control of Saltcedar (*Tamarix*
spp.) to a Nontarget Plant, *Franke-
nia salina. Biological Control* 35:
265–75.

Eberts, Debra. n.d. *Biocontrol
of Saltcedar: Background and
Considerations for Possible Use on
the Canadian River, Texas.* U.S.
Bureau of Reclamation, Technical Service Center Environmental Applications and Research
Group. Technical Memorandum
No. 8220–01–1. http://www.usbr.
gov/pmts/eco_research/011.html
(accessed June 15, 2007).

Echelle, A. A., C. Hubbs, and
A. F. Echelle. 1972. Develop-
mental Rates and Tolerances of
the Red River Pupfish,
Cyprinodon rubrofluviatilis.

Southwestern Naturalist 17(1):
55–60.

Flores, Dan. 1990. *Caprock Can-
yonlands: Journeys into the Heart
of the Southern Plains.* Austin:
University of Texas Press.

Hart, Charles, Larry D. White,
Alyson McDonald McDonald,
and Zhuping Sheng. 2005. Saltcedar Control and Water Salvage on
the Pecos River, Texas. *Journal
of Environmental Management*
74(4): 399–409.

Hendrickson, Kenneth E., Jr. 2008.
Brazos River. *Handbook of Texas
Online.* http://www.tshaonline
.org/handbook/online/articles/
BB/rnb7.html (accessed June 21,
2008).

Hickerson, Nancy P. 2009. Jumano
Indians. *Handbook of Texas Online.* http://www.tshaonline.org/
handbook/online/articles/JJ/bmj7
.html (accessed July 15, 2009).

High Plains Water Conservation
District No. 1. 2008. The Ogallala Aquifer. http://www.hpwd.
com/the_ogallala.asp (accessed
July 15, 2008).

Hunt, William R. 2009a. Eliasville,
Texas. *Handbook of Texas Online.*
http://www.tshaonline
.org/handbook/online/articles/
EE/hle11.html (accessed August
24, 2009).

———. 2009b. South Bend, Texas.
Handbook of Texas Online. http://
www.tshaonline.org/handbook/
online/articles/SS/hls63.html
(accessed August 24, 2009).

Lipscomb, Carol A. 2009. Comanche Indians. *Handbook of Texas Online.* http://www.tshaonline.org/handbook/online/articles/CC/bmc72.html (accessed July 15, 2009).

Matthews, Sallie Reynolds. 2005 [1936]. *Interwoven: A Pioneer Chronicle.* College Station: Texas A&M University Press.

May, Janice C. 2008. Government. *Handbook of Texas Online.* http://www.tshaonline.org/handbook/online/articles/GG/mzgfq.html (accessed May 13, 2008).

Renfro, J. Larry, and Loren G. Hill. 1970. Factors Influencing the Aerial Breathing and Metabolism of Gars (*Lepisosteus*). *Southwestern Naturalist* 15(1): 45–54.

Stephens, A. Ray, and William M. Holmes. 1988. *Historical Atlas of Texas.* Norman: University of Oklahoma Press.

Texas Beyond History. 2009a. Fort Griffin. http://www.texasbeyondhistory.net/forts/griffin/index.html (accessed June 17, 2009)

———. 2009b. The Most Dangerous Prairie in Texas. http://www.texasbeyondhistory.net/forts/griffin/prairie.html (accessed June 17, 2009)

———. 2009c. The Post on Government Hill. http://www.texasbeyondhistory.net/forts/griffin/post.html (accessed June 17, 2009)

———. 2010. Alibates Flint Quarries and Ruins. http://www.texasbeyondhistory.net/alibates/index.html (accessed December 15, 2010).

Texas Parks and Wildlife Department 2008a. Alan Henry Reservoir. http://www.tpwd.state.tx.us/fishboat/fish/recreational/lakes/alan_henry/

Texas Parks and Wildlife Department 2008b. Water Body Records for Brazos River. http://www.tpwd.state.tx.us/fishboat/fish/action/waterecords.php?WB_code=1062

Texas Water Development Board. 2008. Seymour Aquifer. http://www.twdb.state.tx.us/publications/reports/GroundWater Reports/GWReports/R345%20Aquifers%200f%20Texas/Majors/seymour.pdf (accessed June 21, 2008).

U.S. Army Corps of Engineers. 2007. *Diorhabda elongata* Brulle—"Salt Cedar Leaf Beetle." http://el.erdc.usace.army.mil/pmis/biocontrol/html/diorhabd.html (accessed August 1, 2007).

U.S. Census Bureau. 2009a. Population of Counties by Decennial Census: 1900 to 1990. http://www.census.gov/population/cencounts/tx190090.txt (accessed June 17, 2009).

———. 2009b. State and County Quick Facts. http://quickfacts.census.gov/qfd/states/48/48059.html (accessed June 17, 2009).

U.S. Environmental Protection Agency. 2007. Drinking Water State Revolving Fund Program. Office of Ground Water and Drinking Water. EPA 816-R-07–002. Washington, D.C.: U.S. Environmental Protection Agency. www.epa.gov/safewater/dwsrf (accessed July 12, 2008).

U.S. Forest Service. 2008. Tamerix spp. http://www.fs.fed.us/database/feis/plants/tree/tamspp/all.html#TAXONOMY (accessed July 15, 2008).

Wetzel, Robert G. 2001. *Limnology: Lake and River Ecosystems.* 3rd ed. San Diego, Calif.: Academic Press.

Young, W. J., and R. T. Kingsford. 2006. Flow Variability in Large Unregulated Dryland Rivers. In *Ecology of Desert Rivers,* edited Richard Kingsford, 11–46. Cambridge: Cambridge University Press.

Chapter 5: John Graves's Dammed River

Armbruster, Henry C. 2005. Torrey Trading Houses. *Handbook of Texas Online.* http://www.tsha.utexas.edu/handbook/online/articles/TT/dft2.html (accessed December 23, 2005).

Bestplaces.net. 2009. Granbury, Texas. http://www.bestplaces.net/city/profileaspx?ccity=&city=Granbury_TX

Bozka, Larry. 2007. Stealth Fishing. *Texas Parks and Wildlife Magazine,* June. http://www.tpwmaga

zine.com/archive/2007/jun/ed_1/ (accessed July 18, 2009).

Brazos River Authority. 2007. Planning for the "Great Storms." http://www.brazos.org/Newsletter/Fall_2007_Rain.asp (accessed January 12, 2008).

Brockman, John Martin. 2010. Port Sullivan, Texas. *Handbook of Texas Online*. http://www.tshaonline.org/handbook/online/articles/hrp52 (accessed December 10, 2010).

Burnett, Jonathan. 2008. *Flash Floods in Texas*. College Station: Texas A&M University Press.

Campbell, Randolph B. 2003. *Gone to Texas: A History of the Lone Star State*. New York: Oxford University Press.

Center for Research in Water Resources. 1997. Surface Water Balance. http://www.crwr .utexas.edu/reports/1997/ rpt97-1/SECT5.HTM

City-data. 2009. Glen Rose, Texas, Population. http://www.city-data .com/city/Glen-Rose-Texas.html (accessed July 15, 2009).

City of Waco. 2007. Lake Brazos. http://www.wacowater.com/lake-brazos-dam.html

Collins, Michael B. 2004. Archeology in Central Texas. In *The Prehistory of Texas*, edited by Timothy K. Perttula, 101–126. College Station: Texas A&M University Press.

Committee on Flood Control. 1919. Brazos and Colorado Rivers in

Texas. Hearings before the Committee on Flood Control of the House of Representatives, Sixty-fifth Cong., Third sess., Tuesday, February 26, 1919. Washington, D.C.: Government Printing Office. Kessinger Publishing Legacy Reprints.

Conger, Roger. 1964. *A Pictorial History of Waco*. Waco: Texian Press.

———. 2006. Waco, Texas. *Handbook of Texas Online*. http://www. tsha.utexas.edu/handbook/online/ articles/WW/hdw1.html (accessed January 3, 2006).

Cooke, G. Dennis, Eugene B. Welch, Spencer A. Peterson, and Peter R. Newroth. 1986. *Lake and Reservoir Restoration*. Boston: Butterworths.

Cordell, H. Ken. 2004. *Outdoor Recreation for 21st Century America: A Report to the Nation; The National Survey on Recreation and the Environment*. State College, Pa.: Venture Publishing.

Dunn, David D., and Timothy H. Raines. 2001. *Indications and Potential Sources of Change in Sand Transport in the Brazos River, Texas*. Water-Resources Investigations Report 01–4057. Austin, Tex.: U.S. Geological Survey.

Ferrer, Ada. 2010. Glen Rose, Texas. *Handbook of Texas Online*. http://www.tshaonline.org/ handbook/online/articles/hjg03. (accessed December 12, 2010).

Foster, William C. 1995. *Spanish Expeditions into Texas*, 1689–

1768. Austin: University of Texas Press.

Grahame, Kenneth. 2003 [1908]. *The Wind in the Willows*. Cambridge, Mass.: Candlewick Press.

Graves, John. 2002 [1960]. *Goodbye to a River*. New York: Knopf. Reprint, New York: Vintage Books.

Handbook of Texas Online. 2009. Brazos, Texas. http://www.tshaonline.org/ handbook/online/articles/BB/hnb74 .html (accessed July 28, 2009).

Hendrickson, Kenneth E., Jr. 1981. *The Waters of the Brazos: A History of the Brazos River Authority, 1929–1979*. Waco: Texian Press.

Jobin, William. 1998. *Sustainable Management for Dams and Waters*. Boca Raton, Fla.: Lewis Publishers.

La Vere, David. 2004. *Texas Indians*. College Station: Texas A&M University Press.

Lockwood & Andrews. 1955. Review of Certain Phases of the Report on "Plan Development for Water Conservation Flood Control and Power Brazos River, Texas." Prepared for the Brazos River Authority, Mineral Wells, Tex.

Mayborn, Ted W. 2010. Granbury, Texas. *Handbook of Texas Online*. http://www.tshaonline.org/ handbook/online/articles/hgg03 (accessed December 12, 2010).

Minor, David. 2010. Dennis, Texas. *Handbook of Texas Online*. http:// www.tshaonline.org/handbook/

online/articles/hnd11 (accessed December 9, 2010).

Morgan, Tiffany. 2006. *Evaluation of Potential Control Mechanism for Golden Algae Control in the Upper Brazos River Basin. Final Report*. U.S. Environmemntal Protection Agency Grant Number CP-976859-01-0. Waco, Tex.: Brazos River Authority.

Murlin, Bill, ed. 1991. *Woody Guthrie: Roll on Columbia; The Columbia River Collection*. Bethlehem, Pa.: Sing Out Publications.

Myres, Sandra L. 2010. Fort Graham. *Handbook of Texas Online*. http://www.tshaonline.org/handbook/online/articles/qbf21 (accessed December 10, 2010).

National Park Service. 2009. *Nationwide River Survey, Texas Segments*. http://www.nps.gov/ncrc/programs/rtca/nri/states/tx.html (accessed July 30, 2009).

Odintz, Mark. 2010. Stone City, Texas. *Handbook of Texas Online*. http://www.tshaonline.org/handbook/online/articles/hrs68 (accessed December 10, 2010).

Puryear, Pamela Ashworth, and Nath Winfield Jr. 1976. *Sandbars and Sternwheelers: Steam Navigation on the Brazos*. College Station: Texas A&M University Press.

Roemer, Ferdinand. 1935 [1849]. *Texas; with Particular Reference to German Immigration and the Physical Appearance of the Country*. Translated by Oswald Mueller.

San Antonio: Standard Printing Company.

Scarbrough, Linda. 2005. *Road, River, and Ol' Boy Politics*. Austin: Texas State Historical Association.

Skovgaard, Alf, and Per Juel Hansen. 2003. Food Uptake in the Harmful Alga *Prymnesium parvum* Mediated by Excreted Toxins. *Limnology and Oceanography* 48(3): 1161–1166.

Texas Parks and Wildlife Department. 2008. Striped Bass (*Morone saxatilis*). http://www.tpwd.state.tx.us/huntwild/wild/species/str/

Texas Parks and Wildlife Department. 2010a. Stocking Rates for Striped Bass. http://www.tpwd.state.tx.us/fishboat/fish/action/stock_byspecies.php?timeframe=selectyear&species=0111&year=2009&Submit=Go (accessed July 18, 2010).

Texas Parks and Wildlife Department. 2010b. Biology of Golden Alga, *Prymnesium parvum*. http://www.tpwd.state.tx.us/landwater/water/environconcerns/hab/ga/bio.phtml

Titre, John, Chris Jones, Justin Woods, and Memory Karamozondo. 2006. *Boating Capacity Study: Possum Kingdom Lake, Texas*. Report to the Brazos River Authority.

Trungale Engineering & Science. 2007. *Instream Flow Needs for the Brazos River near Glen Rose, Texas*. Prepared for Friends of the

Brazos River. http://www.friendsofthebrazos.org/archive/InstreamFlowNeedsAnalysis-Fina17–2007.pdf (accessed July 21, 2010).

U.S. Army Corps of Engineers. 2008. History of Waco Lake. http://www.swf-wc.usace.army.mil/waco/Information/History.asp (accessed June 12, 2008).

U.S. Census Bureau. 2009. Population of Counties by Decennial Census: 1900 to 1990. http://www.census.gov/population/cencounts/tx190090.txt

U.S. Department of the Interior, Fish and Wildlife Service, and U.S. Department of Commerce, U.S. Census Bureau. 2006. *2006 National Survey of Fishing, Hunting, and Wildlife-Associated Recreation*. http://www.census.gov/prod/www/abs/fishing.html (accessed July 18, 2008).

U.S. Geological Survey. 2008. National Water Information System Web Interface Peak Streamflow for the Nation USGS 08096500 Brazos Rv at Waco, TX. http://nwis.waterdata.usgs.gov/nwis/peak?siteno=08096500&agency_cd=USGS&format=html (accessed December 3, 2008).

Wetzel, Robert G. 2001. *Limnology: Lake and River Ecosystems*. 3rd ed. San Diego, Calif.: Academic Press.

Willingham, John. 2010. Barnard, George. *Handbook of Texas Online*. http://www.tshaonline.org/

handbook/online/articles/fba68 (accessed December 8, 2010).

World Commission on Dams. 2000. *Dams and Development: A New Framework for Decision-Making*. London: Earthscan Publications.

Youcha, Joe. 1998. Why Start a Community Boatbuilding Program? In *Community Boatbuilding Manual*, 4–7. Brooklin, Me.: WoodenBoat Publications.

Chapter 6: The (Almost) Free Brazos

Anhaiser, Bettye J. 2009. Sugar Land, Texas. *Handbook of Texas Online*. http://www.tshaonline.org/handbook/online/articles/SS/hfs10.html (accessed August 3, 2009).

Best Places. 2009a. San Felipe, Texas. http://www.bestplaces.net/city/San_Felipe-Texas.aspx (accessed July 23, 2009).

———. 2009b. Richmond, Texas. http://www.bestplaces.net/City/Richmond-Texas.aspx (accessed July 23, 2009).

Best Places. 2009c. Rosenberg, Texas. http://www.bestplaces.net/city/Rosenberg-Texas.aspx (accessed July 23, 2009).

Bishop, Curtis. 2009. McKinney, Williams and Company. *Handbook of Texas Online*. http://www.tshaonline.org/handbook/online/articles/MM/dfm1.html (accessed August 6, 2009).

Bonner, Timothy, and Dennis T. Runyan. 2007. *Fish Assemblage Changes in Three Western Gulf Slope Drainages*. Final Project Report (2005–483–033) Submitted to Texas Water Development Board.

Brazosport Area Chamber of Commerce. 2009. http://www.brazosport.org/ (accessed July 15, 2009).

Britton, Karen Gerhardt, Fred C. Elliott, and E. A. Miller. 2008. Cotton Culture. *Handbook of Texas Online*. http://www.tshaonline.org/handbook/online/articles/CC/afc3.html (accessed July 25, 2008).

Cabeza de Vaca, Alvar Núñez. 2002. *Chronicle of the Narváez Expedition*. Translated by Fanny Bandelier. New York: Penguin Books.

Campbell, Randolph B. 2003. *Gone to Texas: A History of the Lone Star State*. New York: Oxford University Press.

Christian, Carole E. 2009. Washington on the Brazos. *Handbook of Texas Online*. http://www.tshaonline.org/handbook/online/articles/WW/hvw10.html (accessed August 3, 2009).

City-data.com. 2009a. West Columbia, Texas. http://www.city-data.com/city/West-Columbia-Texas.html (accessed July 17, 2009).

———. 2009b. Brazoria, Texas. http://www.city-data.com/city/Brazoria-Texas.html (accessed July 17, 2009).

City of Sugar Land. 2007. The Brazos River Park Takes Off at Sugar Land Memorial Park Grand Opening. http://www.sugarlandtx.gov/tools/np/program/view.asp?ID=11623

Cockrell, Jay. 2005. Texas Rice Growers Cited for Stewards of Economy and Environment. *Southwest Farm Press*, February 1.

Crawford, Ann Fears. 2009. Fulshear, Texas. *Handbook of Texas Online*. http://www.tshaonline.org/handbook/online/articles/FF/hlf34.html (accessed August 10, 2009).

Dethloff, Henry C. 2009. Rice Culture. *Handbook of Texas Online*. http://www.tshaonline.org/handbook/online/articles/RR/afr1.html (accessed August 5, 2009).

Dienst, Alex. 1909. "The Navy of the Republic of Texas." *Quarterly of the Texas State Historical Association*, 12, no. 4, pp. 249–75, digital images, ttp://texashistory.unt.edu/ark:/67531/metapth101048 (accessed December 15, 2010), University of North Texas Libraries, The Portal to Texas History, http://texashistory.unt.edu; crediting Texas State Historical Association, Denton, Texas, http://texashistory.unt.edu/ark:/67531/metapth101048/m1/311/

Dunn, David D., and Timothy H. Raines. 2001. *Indications and Potential Sources of Change in Sand Transport in the Brazos River, Texas*. Water-Resources Investigations Report 01–4057. Austin, Tex.: U.S. Geological Survey.

Few, Joan. 2007. *Sugar, Planters, Slaves, and Convicts: The History and Archaeology of the Lake Jackson Plantation, Brazoria County, Texas*. Gold Hill, Colo.: Few Publications.

FishQuest 2010. Alligator Gar Trek. http://www.fishquest.com/Quest .asp?Option=TripDetail&Detail= 115 (accessed March 15, 2010).

Hallstein, Anna. 2009. Brazoria, Texas. *Handbook of Texas Online*. http://www.tshaonline.org/hand book/online/articles/BB/hgb10. html (accessed August 3, 2009).

Jackson, Charles Christopher. 2009. San Felipe de Austin. *Handbook of Texas Online*. http://www .tshaonline.org/handbook/online/ articles/SS/hls10.html (accessed August 3, 2009).

Jones, Marie Beth. 2009a. Quintana, Texas. *Handbook of Texas Online*. http://www.tshaonline.org/hand book/online/articles/QQ/hnq3. html (accessed August 3, 2009).

———. 2009b. Varner-Hogg Plantation State Historical Park. *Handbook of Texas Online*. http://www .tshaonline.org/handbook/online/ articles/VV/ghv1.html (accessed August 3, 2009).

Jordan, Terry G. 1984. *Texas: A Geography*. With John L. Bean Jr. and William M. Holmes. Boulder, Colo.: Westview Press.

Kelly, Sean M. 2010. *Los Brazos de Dios: A Plantation Society in the Texas Borderlands*, 1821–1865. Baton Rouge: Louisiana State University Press.

Kleiner, Diana J. 2005. Freeport, Texas. *Handbook of Texas Online*. http://www.tsha.utexas.edu/ handbook/online/articles/FF/ hef3.html (accessed December 23, 2005).

———. 2009a. Brazoria County. *Handbook of Texas Online*. http:// www.tshaonline.org/handbook/ online/articles/BB/hcb12.html (accessed August 5, 2009).

———. 2009b. Brazosport, Texas. *Handbook of Texas Online*. http:// www.tshaonline.org/handbook/ online/articles/BB/hdb3.html (accessed August 3, 2009).

———. 2009c. Imperial Sugar Company. *Handbook of Texas Online*. http://www.tshaonline.org/hand book/online/articles/II/diicy.html (accessed August 4, 2009).

La Vere, David. 2004. T*exas Indians*. College Station: Texas A&M University Press.

Leffler, John. 2009. Richmond, Texas. *Handbook of Texas Online*. http://www.tshaonline.org/hand book/online/articles/RR/hfr4. html (accessed August 3, 2009).

Lucko, Paul M. 2009. Prison System. *Handbook of Texas Online*. http://www.tshaonline.org/hand book/online/articles/PP/jjp3.html (accessed August 20, 2009).

Mooney, Kevin E. 1998. Texas Centennial 1936: Music and Identity. PhD dissertation, University of Texas at Austin.

Moore, Richard W. 2009. Baron de Bastrop. *Handbook of Texas Online*. http://www.tshaonline .org/handbook/online/articles/ BB/fbaae.html (accessed August 4, 2009).

Myers, Lynda Jill. 2009. Rosenberg, Texas. *Handbook of Texas Online*. http://www.tshaonline.org/hand book/online/articles/RR/her2. html (accessed August 4, 2009).

Nagle, C. 1910. *Irrigation in Texas*. Washington, D.C.: Government Printing Office.

National Oceanographic and Atmospheric Administration. 2010a. National Weather Service Forecast Office. College Station Extremes, Normals and Annual Summaries. http://www.srh.noaa. gov/hgx/?n=climate_cll_normals_ summary (accessed December 15, 2010).

———. 2010b. National Weather Service Forecast Office. Galveston Extremes, Normals and Annual Sum// www.srh.noaa.gov/hgx/?n=climate_ gls_normals_summary (accessed December 15, 2010).

Osting, Tim, Jordan Furnans, and Ray Mathews. 2004. *Surface Connectivity between Six Oxbow Lakes and the Brazos River, Texas*. Austin: Surface Water Resources Division Texas Water Development Board.

Osting, Tim, Ray Mathews, and Barney Austin. 2004. *Analysis of*

Instream Flows for the Lower Brazos River—Hydrology, Hydraulics, and Fish Habitat Utilization. Final Report. Vol. 1—Main Report. Austin: Surface Water Resources Division, Texas Water Development Board.

Port of Freeport. 2009. http://www.portfreeport.com/about.htm

Puryear, Pamela Ashworth, and Nath Winfield Jr. 1976. *Sandbars and Sternwheelers: Steam Navigation on the Brazos*. College Station: Texas A&M University Press.

Rice, Margaret R. 2009. Lake Jackson, Texas. *Handbook of Texas Online*. http://www.tshaonline.org/handbook/online/articles/LL/he13.html (accessed August 3, 2009).

Ricklis, Robert A. 1996. *The Karankawa Indians of Texas: An Ecological Study of Cultural Tradition and Change*. Austin: University of Texas Press.

———. 2004. Prehistoric Occupation of the Central and Lower Texas Coast. In *The Prehistory of Texas*, edited by Timothy K. Perttula, 155–80. College Station: Texas A&M University Press.

Roemer, Ferdinand. 1935 [1849]. *Texas; with Particular Reference to German Immigration and the Physical Appearance of the Country*. Translated by Oswald Mueller. San Antonio: Standard Printing Company.

Salvant, J. U. 1999. *The Historic Seacoast of Texas* / paintings by J.U. Salvant; essays by David G. McComb. Austin: University of Texas Press.

Scarbrough, Linda. 2005. *Road, River, and Ol' Boy Politics*. Austin: Texas State Historical Association.

Spearing, Darwin. 1991. *Roadside Geology of Texas*. Missoula, Mt.: Mountain Press Publishing.

Steele, D. Gentry, and M. Jimmie Killingsworth. 2007. *Reflections of the Brazos Valley*. College Station: Texas A&M University Press.

Taylor, Thomas Ulvan. 1902. Rice Irrigation in Texas. *Bulletin of the University of Texas*, No. 16. Austin: University of Texas.

Texas Beyond History. 2009. Making Sugar in Nineteenth-Century Texas. http://www.texasbeyondhistory.net/jackson/sugar.html (accessed June 23, 2009).

Texas Parks and Wildlife Department. 2009. Red Shiner (*Cyprinella lutrensis*). http://www.tpwd.state.tx.us/huntwild/wild/species/redshiner/ (accessed March 22, 2010).

———. 2010a. All-tackle Records for the Brazos River. http://www.tpwd.state.tx.us/fishboat/fish/action/alltackle.php?WB_code=1062

———. 2010b. Alligator Gar (*Atractosteus spatula*). http://www.tpwd.state.tx.us/huntwild/wild/species/alg/

———. n.d. Distribution of American

Alligators in Texas. http://www.tpwd.state.tx.us/publications/pwdpubs/media/pwd_lf_w7000_0162.pdf

Thomas, Chad, Timothy H. Bonner, and Bobby G. Whiteside. 2007. *Freshwater Fishes of Texas: A Field Guide*. College Station: Texas A&M University Press.

Townsend, Christi. 2007. A Geographic History of the Brazos River Diversion at Freeport, Texas, and Its Influence on the Development of Freeport Harbor and the Brazosport Region. Master's thesis, Texas State University–San Marcos.

U.S. Bureau of the Census. 2009a. Texas Quick Facts: Lake Jackson. http://quickfacts.census.gov/qfd/states/48/4840588.html (accessed June 23, 2009).

———. 2009b. Texas Quick Facts: Sugar Land. http://quickfacts.census.gov/qfd/states/48/4870808.html (accessed June 23, 2009).

U.S. Coast Guard. 2009. Freeport Station History. http://www.uscg.mil/d8/staFreeport/station-history.asp

Wallis, Jonnie Lockhart. 1967 [1930]. *Sixty Years on the Brazos: The Life and Letters of Dr. John Washington Lockhart*. Los Angeles, Calif.: Privately printed. Reprint, Waco: Texian Press.

Weir, Merle 2009a. East Columbia, Texas. *Handbook of Texas Online*. http://www.tshaonline.org/hand

book/online/articles/EE/hne2
.html (accessed August 3, 2009).
———. 2009b. Old Velasco, Texas. *Handbook of Texas Online.* http://www.tshaonline.org/hand book/online/articles/VV/hvv7. html (accessed August 3, 2009).

Weir, Merle, and Diana J. Kleiner. 2009. West Columbia, Texas. *Handbook of Texas Online.* http://www.tshaonline.org/handbook/online/articles/WW/hgw3.html (accessed August 3, 2009).

Werner, George C. 2009. Gulf, Colorado and Santa Fe Railway Company. *Handbook of Texas Online.* http://www.tshaonline.org/handbook/online/articles/GG/eqg25.html (accessed August 4, 2009).

Zeug, Steven C., Kirk O. Winemiller, and Soner Tarim. 2005. Response of Brazos River Oxbow Fish Assemblages to Patterns of Hydrologic Connectivity and Environmental Variability. *Transactions of the American Fisheries Society,* 134:1389–99.

Chapter 7: The Evolving Brazos

Berry, Thomas. 2006. *Evening Thoughts: Reflecting on Earth as Sacred Community.* San Francisco: Sierra Club Books.

Brazos River Authority. 2002. Questions and Answers for Proposed Allen's Creek Reservoir. http://www.brazos.org/ORGANIZA TION/news_releases/02–4–23_ Allens_Creek_Q&A_1.htm

Osting, Tim, Ray Mathews, and Barney Austin. 2004. *Analysis of Instream Flows for the Lower Brazos River—Hydrology, Hydraulics, and Fish Habitat Utilization.* Final Report. Vol. 1—Main Report. Austin, Texas: Surface Water Resources Division, Texas Water Development Board.

Thoreau, Henry David. N.d. *Walking.* Kindle Edition.

Appendix: Plant and Animal Species of the Brazos River

Benke, Arthur C., and Colbert E. Cushing. 2005. Background and Approach. In *Rivers of North America,* edited by Arthur C. Benke and Colbert E. Cushing, 1–18. Burlington, Mass.: Elsevier Academic Press.

Collins, James P., Claude Gascon, and Joseph R. Mendelson. 2007. Foreword to *Amphibian Conservation Action Plan. Proceedings: IUCN/SSC Amphibian Conservation Summit 2005,* edited by Claude Gascon, James P. Collins, Robin D. Moore, Don R. Church, Jeanne E. McKay, and Joesph R. Mendelson. Arlington, Va.: IUCN/World Conservation Union.

HDR Inc. 2001. *Brazos G. Regional Water Plan.*

Dahm, Clifford, Robert J. Edwards, and Frances P. Gelwick. 2005. Gulf Coast Rivers of the Southwestern United States. In *Rivers of North America,* edited by Arthur C. Benke and Colbert E. Cushing, 181–228. Burlington, Mass.: Elsevier Academic Press.

Rosen, David J., Diane De Steven, and Michael L. Lange. 2008. Conservation Strategies and Vegetation Characterization in the Columbia Bottomlands, and Under-recognized Southern Floodplain Forest Formation. *Natural Areas Journal* 28: 74–82.

Texas Parks and Wildlife Department. 2009. Migration and the Migratory Birds of Texas. http://www.tpwd.state.tx.us/huntwild/wild/birding/migration/ (accessed June 23, 2009).

Vines, Robert A. 1984. *Trees of Central Texas.* Austin: University of Texas Press.

INDEX

Abilene, 66, 71, 75

Abilene State Park, 75

acre-foot, defined, 62

Acrocanthosaurus tracks, 110–11

agriculture: (Amost) Free section, 54, 55, 121, 127–28, 131, 133–34, 140–46; Dammed River section, 83–85, 86, 87*f*, 96; erosion contribution, 26; Google Earth views, 10–11; Lost River section, 50*f*, 51; Many Arms section, 65–66, 67, 68*f*, 71

"Ain't No More Cane on the Brazos" (song), 141

airplane story, 1

Akokisas people, 127

Alan Henry Reservoir, 54–55, 62

algae, golden *(Prymnesium parvum)*, 90

Allen's Creek project, 150–52

alligator gar *(Atractosteus spatula)*, 124

alligators, 124–26, 146, 148*f*

allochthonous aquatic system, in river continuum concept, 29, 31

(Almost) Free section: overview, 115–16; agricultural activity, 121, 127–28, 131, 133–34, 140–46; ecology of, 121–26, 155–60; Google Earth view, 11–12; human history, 126–40, 145; land characteristics, 121; recreation/learning activities, 134, 136–37*f*, 146–49; water characteristics, 9*f*, 116–21, 124

Altithermal, 16

Ambursen Engineering Corporation, 102

America by Rivers (Palmer), 115

American alligator *(Alligator mississippiensis)*, 124–26

amphibian species, list, 158–59

Apache people, 70, 91

aquatic plants, species list, 156–57

aquifers: Brazos River, 35; Ogallala, 51, 54, 62, 65; Seymour, 67

archaeological sites: (Amost) Free section, 55, 57, 126–27, 134; Dammed River section, 90; Lost River section, 51–52, 53*f*; Many Arms section, 70

Army, U.S., 71

Army Corps of Engineers, 41, 97, 98, 103

astronomy, 149

Athabascan people, 70

atmospheric circulation, 12–14. *See also* rainfall patterns

Atractosteus spatula (alligator gar), 124

Austin, Moses, 127–28, 131, 138

Austin, Stephen F., 127–28, 134

Austin's Colony, ix, 127–28, 135, 138, 140

Austin State Park, 12*f*, 146, 147*f*

autochthonous aquatic system, defined, 29

bacteria in hyporheic zone, 37

Barnard, George, 95

Barnard Trading Post, Texas, 95

base flows, in river ecosystems, 32, 34

bass, striped *(Morone saxatilis)*, 88–90

Baylor University, 43, 101*f*

Bell, Josiah, 134

Bell County Museum, 90

Belton Reservoir, 41, 105

Benke, Arthur C., 7

Berry, Thomas, 150

biological productivity. *See* river ecosystems *entries*

Birch Creek, 114

bird species: (Amost) Free section, 28*f*, 122*f*, 142, 147–48*f*; list of, 159; Lost River section, 49, 51, 55, 56*f*; Many Arms section, 67

Blackland Prairie ecoregion, 83, 121, 122*f*

Blackwater Draw, 45, 47*f*, 51–52, 53*f*

boating activities, 106–108

boat song, 128

Bögel, Philip Hendrik Nering, 128

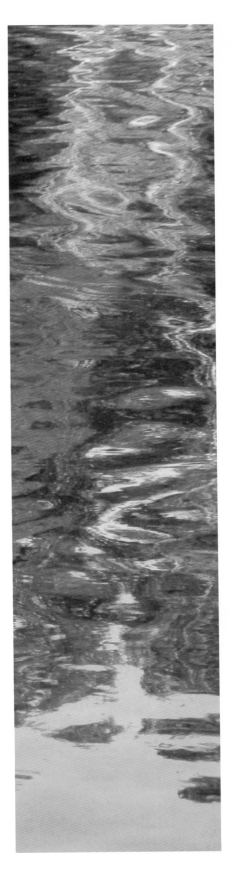

Bomar, George W., 13–14

Bonneville Dam, 105

Bosque Bluffs Paddling Trail, 43

Bosque Museum, 90

Bosque River, 11, 32f, 81, 82, 98

Bozka, Larry, 106–108

braided channels, 17

Brazoria, Texas, 134

Brazoria County, population statistics, 119, 145

Brazos, Texas, 95

Brazos Bend State Park, 146, 148f, 149

Brazos Bridges Paddling Trail, 43

Brazos Canal Company, 129

Brazosport, Texas, 140

Brazos River, overview, ix–x, 1–5, 7–12, 152–54. *See also specific topics, e.g.,* Dammed River section; human history; river ecosystems *entries*

Brazos River Aquifer, 35

Brazos River Authority, 63–64, 102–103, 109, 152

Brazos River Conservation and Reclamation District, 102–103

Brazos River Conservation Coalition, 109

Brazos River Improvement Association, 102

Brazos River Trail, 44

bridges, 10f, 96–97f

buffalo, 54, 71, 72

Butler, Y. M. H., 131

Cabeza de Vaca, 127

caddisflies, 37, 39

Caddo Lake, 24

Caddo people, 91–92

Cameron, William, 98

Cameron Park, 43, 98, 101

Campbell, Randolph, 92

Camp Cooper, 71

Camp Wolters, 110

Canadian River, 54, 70

canal projects, 129, 142

Cane and Rice Belt Irrigation Company, 142

cattle ranching, 54, 71

Cavelier, René-Robert, 92

Cavoque people, 127

cemeteries, prehistoric people, 126

cfs (cubic feet per second), defined, 19

channel bars, 17

Civilian Conservation Corps, 112, 114

Clean Rivers Program, 21, 81, 85t, 126

Clear Fork, 59f, 61, 73–74f

Cleburne State Park, 111–12

climate characteristics, 12–16. *See also* rainfall patterns

Clovis people, 51–52, 53f, 90, 126

coarse particle organic matter (CPOM), defined, 29

Coco people, 93

College Station, Texas, 18, 116, 117

Collins, Michael, 90

Colorado River, 16

Columbia River, 105

Columbus, Christopher, 140

Comanche Peak Nuclear Power Plant, 95

Comanche people, 54, 70–71, 91

Community Boatbuilding Manual, 108

conservation pool, defined, 24, 26

convict labor, 140–41, 142

Coronado, Francisco Vázquez de, 49, 70

Corps of Engineers, 41, 97, 98, 103

cotton production: (Amost) Free section, 128, 131, 140, 141–42; Dammed River section, 96; Many Arms section, 54

cowboy song, 94

CPOM (coarse particle organic matter), defined, 29

crayfish species, 29, 158

Crockett, Elizabeth, 95

Croix, Teodoro de, 92

Croton Creek, 63

crustacean species, 37–39

Crystal Falls, Texas, 72, 74f

cubic feet per second (cfs), defined, 19

Cushing, Colbert E., 7

cut banks, 17, 18f

Cyprinnella butrensis (red shiner), 124

Cyprinodon rubrofluviatilis (Red River pupfish), 67

dairy farming, 82, 86

Dammed River section: overview, 76–77; dam projects/operations, 102–103, 105–106; ecology of, 85–90, 155–60; Google Earth views, 10–11; human history, 90–102; land characteristics, 19, 37, 81, 82–83, 88f; recreation/learning activities, 41, 63, 96, 104f, 106–14; water characteristics, 3, 9f, 25f, 78–82, 83t, 85–88

dams and reservoirs: overview, 23–26, 37; (Amost) Free section, 150–52; Many Arms section, 54–55, 62, 66, 74f. *See also* Dammed River section

Deadose people, 93

DeCordova Bend Dam, 10–11, 82, 95

deep time events, 18–19, 20–21t, 45

deepwater habitat, overview, 37

Dennis, Texas, 95

deposition processes. *See* sediment loads, movements/deposits

detritivores, defined, 37

detritus, defined, 37

Dinosaur Valley State Park, 110–11

Diorhabda elongata Brulle (salt cedar leaf beetle), 70

Double Mountain Fork, 59f, 62

Dow Chemical, 135

droughts, 16, 52, 117

earth movements, 16–19, 21–22t. *See also* geological characteristics, overview; land characteristics

East Columbia, Texas, 134

Ecology of Desert Rivers (Kingsford and Thompson), 45

ecoregions, map, 8f

ecosystems. *See* river ecosystems *entries*

Ecozoic era concept, 150

egrets, 28f

Eliasville, Texas, 72–74

Elm Creek, 75

energy pathways, river ecosystems, 29, 31–34

England, 78–79

erosion processes, 16–19, 67, 117, 120*f.* *See also* sediment loads, movements/ deposits
estuaries, 87–88
European explorers/settlers. *See* human history
evaporation, 87
evapotranspiration, 14–16
Evening Thoughts (Berry), 150

Falls of the Brazos, 28*f*, 97, 152–53*f*
farming. *See* agriculture
ferry operations, 96, 102, 117, 129, 132*f*
filtering collectors, ecological functions, 39
fine particle organic matter (FPOM), defined, 31
fire, 15
fishing, 88–90, 106–108, 124, 136–37*f*
fish species: (Amost) Free section, 124; Dammed River section, 88–90; list of, 158–59; Many Arms section, 35*f*, 67, 69; North America statistics, 29
flooding. *See* water flow
floodplain habitat, overview, 35
flood pool, defined, 24, 26
flood pulse concept, 29, 31, 86, 123–24
Flores, Dan, 62
fluvial geomorphology, defined, 16. *See also* sediment loads, movements/deposits
Folsom people, 52
Fort Belknap, 71, 75
Fort Bend, 133
Fort Bend County, population statistics, 119
Fort Graham, 95–96
Fort Griffin, 71, 72, 73*f*, 75
Fort Griffin State Historic Site, 75
Fort Hubbard, 71
Fort Parker State Park, 112, 114
Fort Phantom Hill, 71
Fort Phantom Hill Reservoir, 66
Fort Sullivan, 101
FPOM (fine particle organic matter), defined, 31
Freeport, Texas, 11, 129, 130*f*, 134–35
French explorers, 92

Friends of the Brazos, 109
Fulshear, Texas, 133

Galveston, Texas, 116, 129, 140
Galveston and Brazos Navigation Company, 138
gar, longnose *(Lepisosteus osseus)*, 67, 69, 124
geological characteristics, overview: deep time events, 18–19, 21*t*, 45; Google Earth view, 10–12; physical geography summarized, 23*t*; water flow effects, 16–18. *See also* land characteristics
Georgetown Reservoir, 41, 105
Giesensechlag, Connie, 86
Giesensechlag, John, 86
Glen Canyon Dam, 82
Glen Rose, Texas, 95
golden algae *(Prymnesium parvum)*, 90
Goodbye to a River (Graves), 1, 3*f*, 105
Google Earth views: (Amost) Free section, 18, 121, 124, 129, 145; Dammed River section, 98; Many Arms section, 64, 65; river length, 8, 10–12
Graham, Texas, 78, 79*f*, 108
Granbury, Texas, 10, 95. *See also* Lake Granbury
Granger Lake, 41, 87, 105
Graves, John, 1, 3*f*, 76, 102, 105, 106
grazer macroinvertebrates, ecological functions, 39
grazing, 54, 71
Griffin, Texas, 72
Grus canadensis (sandhill cranes), 51, 52*f*, 54
Gulf Coast Prairies and Marshes ecoregion, 122*f*
Guthrie, Woodie, 105

habitat types, river ecosystems, 34–37
Han people, 127
Harris County, population statistics, 145
head of slackwater, 37
headwater catchment, 45
high flow pulses, in river ecosystems, 32, 34

Hildalgo Falls, 11, 82, 99*f*, 123*f*
The Historic Seacoast of Texas (McComb), 145
Holley, Mary Austin, 128
Holliday, Vance T., 55
horizontal (reservoir) pool, defined, 24
Horn Shelter site, 90
Houston, Texas, 129, 140, 150–52
Houston and Texas Central Railroad, 131
Houston Chamber of Commerce, 142
Houston Museum of Natural Science, 149
Houston Wilderness, 44
Hubbard Creek Reservoir, 66
human history: (Amost) Free section, 126–40, 145; Dammed River section, 83–85, 90–102; Lost River section, 50*f*, 51–55; Many Arms section, 70–75
hydroelectric power, 24, 26
hyporheic zone, 34–35, 37

impoundments/tanks, 37, 65–66, 72
Indians. *See* human history
insect species, 29, 37–39, 148*f*, 157–58
Interwoven (Matthews), 58
Intracoastal Waterway, 129
invasive species: (Amost) Free section, 124; Dammed River section, 90; ecological impact, 15; list of, 157, 159; Many Arms section, 15, 69–70
invertebrate species, 29, 37–39, 148*f*, 157–58
irrigation. *See* agriculture
Irrigation in Texas (Nagle), 142

Jackson, Mark, 86
Jackson, Penne, 86
Jackson Plantation Archaeological Site, 134
John Graves Scenic Riverway, 109
Johnson, Eileen, 55
Jordan, T. C., 95
Jumanos people, 70
juniper, 15

Karankawa people, 92, 127
Katie Ross, 97
kayaking, 106–108

Killingsworth, Jimmie, 115
Kimmel, Clark, 55
Kimmel, Turner, 55
Kingsford, R. T., 45
Kiowa people, 70–71

Lake Abilene, 75
Lake Aquilla, 41
Lake Belton, 41, 105
Lake Brazos, 100*f*, 101
Lake Georgetown, 41, 105
Lake Granbury, 10*f*, 82, 89, 93, 103, 105
Lake Jackson, Texas (town), 134
Lake Meredith, 54
Lake Meridian, 112
Lake Mineral Wells State Park and Trail-
    way, 110
Lake Powell, 82
Lake Somerville (and park), 41, 114
lakes *vs.* reservoirs, definitions, 24. *See also*
    Dammed River section
Lake Texoma, 89
Lake Waco, 41, 42*f*, 43, 98, 103. *See also*
    Waco, Texas
Lake Whitney (and dam): fish stocking,
    89; flood pool, 26, 103; geological
    characteristics, 20*f*, 82, 89*f*; Google
    Earth views, 11*f*; habitat types, 36*f*, 37;
    recreation/learning areas, 41, 96, 112;
    size of, 37
Lake Whitney State Park, 112
Lampasas River, 105
land characteristics: (Amost) Free sec-
    tion, 121; Dammed River section, 81,
    82–83, 88*f*; Lost River section, 47*f*,
    49–51; Many Arms section, 66–67. *See
    also* geological characteristics, overview
La Salle, René Robert Cavelier, Sieur de,
    92, 126
La Vere, David, 90–91
leaf beetle, salt cedar *(Diorhabda elongata
    Brulle)*, 70
learning activities. *See* recreation/learning
    activities
Lee, Robert E., 71
Leon River, 80*f*, 105, 112

Libertador, 138–39
limestone cliffs, 19, 82*f*, 110
limnologists, defined, 29
Little Brazos River, 81
Little River, 11, 81, 105
littoral areas, 37
Lizzie Fisher, 97
Llano Estacado (Staked Plain), name ori-
    gins, 49. *See also* Lost River section
local biological production. *See* river eco-
    systems *entries*
Lockhart, John Washington, 117, 131
Lockwood & Andrews, 102–103
longnose gar *(Lepisosteus osseus),* 67, 69
Lost River section: ecology of, 51, 160;
    Google Earth views, 10; human history,
    50*f*, 51–55; land characteristics, 47*f*,
    49–51; recreation/learning activities,
    55–57; water characteristics, 9*f*, 45–49
Lovers' Leap, 32*f*
lower reaches, in river continuum concept,
    29–31
lower river. *See* (Almost) Free section
Lubbock, Texas, 54–55, 62
Lubbock County, population statistics, 54
Lubbock Lake Landmark, 55, 57

macroinvertebrates. *See* insect species
mammal species: overview, 38*f*, 39, 160;
    (Amost) Free section, 55, 57, 148*f*;
    Dammed River section, 94, 101; Many
    Arms section, 54, 55, 67, 71
Many Arms of God section: ecology of,
    67–70, 155–60; human history, 70–75;
    land characteristics, 66–67; recreation/
    learning activities, 75; water character-
    istics, 9*f*, 58–66
maps: (Almost) Free section, 116*f*; Dammed
    River section, 77*f*; ecoregions, 8*f*; Lost
    River section, 46*f*; Many Arms section,
    60*f*; weather patterns, 13–14*f*
Marfa Monster, 1, 2–3*f*, 48
Matagorda Bay, 92
Matthews, Sallie Reynolds, 58
McComb, David, 145
McIntyre, Mande, 96

McKinney, Williams and Company, 135
meandering channels, 17–18, 19*f*, 35, 37
Meridian State Park, 112
Mesa Water, 62
mesquite, 15
Mexico, 93, 128, 138
middle reaches, in river continuum con-
    cept, 29–31
military activity. *See* human history
Mill Creek, 72
Mineral Wells, 10*f*, 110
minnows, 124
missions, 92–93
Morone saxatilis (striped bass), 88–90
Morris Sheppard Dam, 77–78*f*, 82
Mother Neff State Park, 112
muddiness (turbidity) characteristics, 26.
    See also sediment loads, movements/
    deposits
Muleshoe National Wildlife Refuge, 55,
    56*f*
museums, 52, 90, 101, 111, 146
mussel species, 29, 159
Mustang, 130–31

Nagle, C., 142, 145
Nails Creek & Trailway, 114
Native Americans. *See* human history
native plants, overview, 15–16, 155–57.
    See also vegetation patterns
Navarro Mills Lake, 41
Navasota, Texas, 131
Navasota River, 11, 114, 123
navigation projects, 97, 128–29
Neff, Isabella Eleanor, 112
Neff, Pat M., 112
New Mexico, 45, 51–52, 53*f*
New Velasco, Texas, 138, 140
Nichols Crossing, 35*f*
North Bosque River, 81
North Fork of the Double Mountain Fork,
    59*f*

O. C. Ivy Reservoir, 66
O'Brian, Patrick, 139
Ogallala Aquifer, 51, 54, 62, 65

oil industry: (Amost) Free section, 119, 121, 129, 143f; Dammed River section, 83; Many Arms section, 68f, 71

Old Velasco, Texas, 135, 138–40

organic matter, in river continuum concept, 29, 31–32

overbank flows, in river ecosystems, 34

oxbow lakes, 18, 35, 37, 123–24, 124, 157

oxygen content, 32

Oyster Creek, 121

paddling trails, 43

Palmer, Tim, 115

Paluxy River, 11, 95, 110–11, 154f

Parker, Cynthia Ann, 71, 112

parks, state, 12f, 75, 110–14

Peta Nocona, 71

petrochemical industry, 119, 134–35, 145. See also oil industry

photosynthesis: Many Arms section, 67; process of, 14–15; in river ecosystems, 29, 31, 37; wastewater impact, 63

physical geography, overview, 23t. See also geological characteristics, overview; land characteristics

Pickens, T. Boone, 62

Pickle, J. J. "Jake," 105

Plains Village people, 70

plant species. See vegetation patterns

playas, 48–49, 51, 56f

Pleurocoelus tracks, 110–11

point bars, 17

pollution. See water quality

population statistics: (Amost) Free section, 119, 130, 131, 133, 134; Dammed River section, 81, 95; Lower River section, 54; Many Arms section, 66, 71–72, 75; statewide, 145

Port Sullivan, Texas, 101

Possum Kingdom Lake: algae invasion, 90; geological characteristics, 19, 37, 82, 88f; Google Earth views, 10; property values, 108; recreation activities, 63, 89, 106, 107f, 110; water level fluctuations, 25f, 103, 105

Possum Kingdom State Park, 63, 110

Possum Kingdom Water Supply Corporation, 64

Post Oak Savanah ecoregion, 83, 121, 122f

power pool, defined, 24, 26

precipitation. See rainfall patterns

prehistory. See archaeological sites

prison labor, 140–41, 142

Proctor Lake, 41

Prymnesium parvum (golden algae), 90

Quanah Parker, 71

Quintana, Texas, 127, 135, 145

Quintana fort, 127

Rábago y Terán, Felipe de, 93

railroads, 54, 101, 102, 110, 131, 133

rainfall patterns: overview, 12–14, 19, 32; (Amost) Free section, 116–17; Dammed River section, 78–80, 81, 88–89, 103, 105; Lost River section, 45, 48; Many Arms section, 61; Marfa Monster characteristics, 1, 2–3f

ranching, 54, 71. See also agriculture

recreation/learning activities: overview, 39–44; (Amost) Free section, 134, 136–37f, 146–49; Dammed River section, 41, 82, 88–90, 96, 104f, 106–14; Lost River section, 55–57; Many Arms section, 75

Red River, 89

Red River pupfish (Cyprinodon rubrofluviatilis), 67

red shiner (Cyprinnella butrensis), 124

Reflections of the Brazos Valley (Steele and Killingsworth), 115

reptile species: overview, 38f, 39, 158; (Amost) Free section, 124–26, 148f; Dammed River section, 110

reservations, Indian, 71

reservoirs vs. lakes, definitions, 24. See also Dammed River section

resilience, defined, 37

"Rice Irrigation in Texas" (Taylor), 142

rice production, 142, 144f

Richmond, Texas, 117, 133

riparian zone habitat, overview, 35

river continuum concept, 29, 31, 85–86, 121, 123

river ecosystems, generally: biological functions, 37–39; dynamic variability, 27–29; flow concepts, 29–34, 85–87; habitat types, 34–37

river ecosystems of Brazos: (Amost) Free section, 28f, 121–26, 148f; Dammed River section, 31–33f, 36f, 88–90, 94, 101; Lost River section, 51, 55; Many Arms section, 35f, 54, 55, 67, 71; recreation/learning activities, 39–44; species list, 155–60

Rivers of North America (Benke and Cushing), 7

Road, River, and Ol' Boy Politics (Scarbrough), 105

Roemer, Ferdinand, 91–92, 93–94, 115, 131, 133

Rolling Plains. See Many Arms of God section

Rosenberg, Texas, 133

Ross, Lawrence Sullivan, 71, 96

Ross, Shapley, 96

Runnels, Hardin, 92

Running Water Draw, 45, 47f, 48

safety guidelines, river exploration, 41

Saibara, Seito, 142

salinity levels, 63–65, 67, 103

salt cedar (Tamarisk spp.), 15, 69–70

Salt Croton Creek, 63

Salt Fork, 45, 59f, 63, 64f

San Antonio and Aransas Pass Railway, 133

San Antonio River, 92

sandhill cranes (Grus canadensis), 51, 52f, 55

San Felipe de Austin, 131

San Gabriel River, 87, 105

Santa Fe Railway, 54, 133

Scarbrough, Linda, 87, 105

scraper macroinvertebrates, 39

Sea Center Texas, 149

sea levels, changes, 126, 127

sediment loads, movements/deposits: overview, 17–18, 26; (Amost) Free section, 119, 121, 123, 129; Dammed River section, 36f, 37, 86–87, 98, 109; Many Arms section, 67; in river continuum concept, 31–32

Seymour Aquifer, 67

Sherman, William Tecumseh, 71

shredders, ecological functions, 37

Singer, George, 55

Six Dam Plan, 86, 102–103

slackwater, head of, 37

slavery, 128, 140–41, 149

snails species, 29

soil characteristics: (Amost) Free section, 121; Dammed River section, 83; Lost River section, 51; Many Arms section, 60f, 61f, 66–67, 68f, 83

songs, 94, 128, 141

Soto, Hernando De, 70

South Bend, Texas, 61–62, 74

Southern Pacific Railroad, 142

Spanish explorers, 49, 70–71, 91, 92–93, 127, 129

stair-step sand deposits, 36f

Staked Plain (Llano Estacado), name origins, 49. See also Lost River section

state parks, 12f, 41–42, 75, 110–14, 146–49

steamboats, 129, 131

Steele, Gentry, 115

Stephen F. Austin State Park, 12f, 146, 147f

Stillhouse Hollow Lake, 41, 105

stock ponds, 37, 65f, 71

Stone City, Texas, 101–102

Stovall Hot Wells, 74

stream flows. See water flow

striped bass (Morone saxatilis), 88–90

subsistence flows, in river ecosystems, 32, 34t

Sugar Land, Texas, 133–34, 135f, 140–41, 145

Taerniopteryx starki (winter stonefly), 157

Tamarisk spp. (salt cedar), 15, 69–70

Taylor, J. W., 138–39

Taylor, Thomas U., 142

tectonic movements, 16

temperatures, 14–16, 32, 79, 116–17

Texas Commission on Environmental Quality, 81

Texas Historical Commission, 42

Texas Parks and Wildlife Department, 41–42, 57, 89

Texas Tech University, 54, 57

Thames River, 78–79

Thompson, J. R., 45

Thrall, Texas, 81

thunderstorms, 1, 2–3f, 48. See also rainfall patterns

Tonkawa people, 91, 93

topography. See geological characteristics, overview; land characteristics

Torrey brothers, 93–94

total suspended solids, 26

Trinity River, 16

turbidity, defined, 26. See also sediment loads, movements/deposits

turfgrass crops, 86, 142

upper reaches, in river continuum concept, 29, 30f

upper river section. See Lost River section; Many Arms of God section

U.S. Fish and Wildlife Service, 55

Varner-Hogg Plantation State Historic Site, 42, 134, 141f, 149

vegetation patterns: overview, 14–16; (Amost) Free section, 117, 119f, 121–24, 147–48f; Dammed River section, 33f, 83, 84f, 86; Lost River section, 51; Many Arms section, 67, 69–70; species listed, 155–57. See also river ecosystems, generally

Velasco Bar, 129

Velasco communities, 135, 138–40

Vincedor del Alamo, 138–39

Waco, Texas: ecosystems, 31–33f, 42f; flow rates, 21, 78; Google Earth views, 11; historical summary, 96–101; rainfall, 78; recreation/learning activities, 42f, 43; water quality, 26, 82. See also Lake Waco

Washington-on-the-Brazos, 129–31

Washington-on-the-Brazos State Historic Site, 132f, 146

wastewater treatment, 62–63, 121

water flow: (Amost) Free section, 117–21; Dammed River section, 23–24, 78–81, 83t, 103, 105; Lost River section, 9f, 45–49; Many Arms section, 9f, 58–62; peak/low patterns summarized, 19, 21, 23; rainfall effects, 12–14; sediment movement impact, 17–18. See also river ecosystems entries

water quality: overview, 26; (Amost) Free section, 117, 121; Dammed River section, 81–82, 85t, 103; Many Arms section, 62–66

weather patterns, overview, 12–16. See also rainfall patterns; temperatures

West Columbia, Texas, 134

wetlands, 42f, 43, 142, 157

White River, 45

Whitney Dam. See Lake Whitney (and dam)

Wichita people, 91

wildfire, 15

wildlife. See bird species; mammal species; reptile species

wildlife refuges/management areas, 41, 42, 43, 55, 56f, 114

wildness values, conserving, 152–54

Wilson-Leonard site, 90

The Wind in the Willows, 108

winter stonefly (Taerniopteryx starki), 157

Wood, Tillotson, 117

woody debris, 31–32

Yazoo River, 81

Yellow House Draw, 45, 47f

zoo, 43